Beliefs and Facts of Creation

Beliefs and Facts of Creation

Adrian Bjornson

Addison Press
Woburn, Massachusetts
www.olduniverse.com

Published by: **Addison Press**
400 West Cummings Park
PMB 1725-111
Woburn, MA 01801
www.olduniverse.com

Copyright © 2005 by Addison Press

All rights reserved under International and Pan-American Copyright Conventions. No part of this book may be reproduced by any mechanical, photographic, or electronic process, nor may it be stored in a retrieval system, transmitted, or otherwise copied for public or private use, without written permission of Addison Press.

Printed in the United States of America

Publisher's Cataloguing-in-Publication
 (Provided by Quality Books, Inc.)

   Bjornson, Adrian.
     Beliefs and facts of creation / Adrian Bjornson.
     p. cm.
     Includes bibliographical references and index.
     ISBN 0-9703231-4-X

     1. Cosmology--Popular works. 2. Creation.
  I. Title.

QB982.B55 2005   523.1
                           QBI05-200006

Dedication

This book is dedicated to

Professor Huseyin Yilmaz

*Who has devoted his life to develop his theory of gravity,
which extends the Einstein relativity concepts
to achieve the goal that Albert Einstein sought.*

Acknowledgements

I am grateful for the patient assistance that Prof. Huseyin Yilmaz has provided in explaining his *Theory of Gravity* and the principles of the Einstein *General Theory of Relativity*.

I thank Dr. William C. Keel of the University of Alabama for his excellent photograph of the M51 Whirlpool Galaxy with its companion galaxy NGC 5195. This was taken on the 1.1-meter Hall telescope at the Lowell Observatory. I have used this photograph on the front cover and in Figure 1-1.

Foreword

Organization of the Book

This book is a simplified version of earlier documents by the author that have addressed the scientific aspects of the mystery of Creation. To obtain more details, the reader is referred to References [1, 2, 4] listed in the *Bibliography* at the end of the book. The following is an abbreviated list of these references:

Believe [1]:   *A Universe that We Can Believe*
Story [2]:   *The Scientific Story of Creation*
Website [4]:   *Internet* website, **www.olduniverse.com**

In this book, these three documents are referred to as *Believe* [1], *Story* [2], and *Website* [4]. The *Website* [4] shows how the other books can be purchased.

This book is sufficiently simple to be understood by the average reader, yet it deals with profound scientific issues. Five appendices give technical details to support the material in the body of the book.

At the end of the book is a *Glossary* that defines terms used in the book. The book has two sets of bibliographic references, which are expressed in brackets such as [10]. Those with the prefix Y, such as [Y10], refer to scientific papers on the Yilmaz theory written by Professors Huseyin Yilmaz and Carroll O. Alley.

Theme of the Book

Since the beginning of historical time, humans have pondered the mystery of Creation. Science has discovered a great deal about this great mystery, but many of the scientific claims of astronomers are confused.

There is strong evidence that all stars, including our sun, were created from huge clouds of hydrogen gas. Heavy elements are created from hydrogen in nuclear reactions that occur when a massive star explodes as a supernova. These heavy elements are picked up in the hydrogen clouds that create new stars, and are the solid matter that forms planets like earth. The book explains the creation of our sun and our earth more than 4 billion years ago, and the creation of life on earth.

There is firm evidence supporting these aspects of Creation, but the process that created the hydrogen clouds remains a mystery.

Astronomers insist that our whole universe began as a "singularity" of infinitesimal size that exploded with a Big Bang about 15 billion years ago. This conclusion was derived from computer studies of the Einstein relativity theory, but *Einstein considered all singularity predictions derived from his theory to be physically impossible.*

Astronomers claim that black holes exist; yet the star inside a black hole must be squeezed into a singularity of infinite density.

An alternate approach to this mystery is provided by the Yilmaz theory of gravity, first published in the prestigious *Physical Review* in 1958. The Yilmaz theory applies the principles of general relativity, and so is a refinement of the Einstein theory. The Yilmaz theory does not allow a singularity, and predicts a Steady State universe of infinite age.

Since the Yilmaz theory is a refinement of the Einstein theory, it demonstrates that the principles of general relativity are inconsistent with the physically impossible Big Bang and black hole singularities. This agrees with Einstein's insistence that "singularities do not exist in physical reality".

Despite the great mathematical complexity of the Einstein theory, this book shows that the principles of the Einstein and Yilmaz theories can be explained in a simple manner that can be readily understood by the average reader. This should give readers the insight to decide what scientific claims they should believe about Creation, and when to conclude that scientists are confusing beliefs with facts.

Mathematical Details in the Website

The *Website* [4] ***www.olduniverse.com****.*is available at no cost on the internet. The *Addendum* section of the *Website* (Page 5) gives a detailed analysis of the Yilmaz theory and is directed toward readers with scientific training. This mathematical analyses demonstrates that the Yilmaz theory has a profound mathematical foundation.

Contents

Dedication and acknowledgements	*v*
Foreword	*vi*
Preface	*xiii*
1. Introduction	**1**
Our search for the truth	1
Our mystical Milky Way	2
Creation of our sun and our earth	3
Nuclear fusion and nuclear fission	*3*
Formation of our sun and the stars	4
Formation of our solar system	*5*
Creation of the earth	*6*
Creation of life on earth	*7*
Darwin's theory of evolution	8
Principle of natural selection	*8*
Objections to the Darwin theory	*9*
The DNA evidence	*10*
The emergence of human awareness	11
How life evolved on earth	12
2. The creation of life on earth	**13**
Early microscopic life	13
Summary of the evolution of life on earth	15
The first animals	17
Severe climates of the young earth	*17*
Development of the fish	19
Amphibians invade the land	20
Spread of plants over the land	22
The reign of the reptiles	23
The Synapsids	*24*
The Dinosaurs	*25*
The slow rise of the mammals	26
The ascent of humans	27
Australopithecus and the Homo genus	*27*
Hunting capabilities of early man	*28*
From Homo Erectus to anatomically modern humans	*29*
The development of sophisticated language	*30*
How the Homo genus evolved	*32*
Was there a divine spark in the development of humanity?	*34*
The growth of civilization	*35*
Beyond the earth	35
Geological periods	35

3. Historical foundation of astronomy — 36
- Ancient astronomical observations — 36
- The celestial sphere of the stars — 36
  - *The ecliptic path followed by the sun* — *38*
  - *Development of our calendar* — *40*
  - *The path followed by the moon* — *41*
  - *The wandering stars* — *41*
  - *The crystalline spheres of Pythagoras and Aristotle* — *42*
  - *The spherical earth* — *42*
- The Almagest of Ptolemy — 43
- Popular acceptance of the spherical earth — 45
- The Copernicus revolution in astronomy — 46
- The planetary laws of Kepler — 47
- Galileo's telescope observations of the heavens — 48
- Galileo's measurement of falling bodies — 49
- The development of the telescope — 50
- Newton's theory of gravity — 51
  - *How Cavendish weighed the earth* — *53*
  - *Orbit of the earth around the sun* — *54*
  - *Why are astronauts weightless?* — *56*
- The independent invention of calculus by Leibniz — 57
- Newton's revolutionary research on optics — 57

4. The solar system and the stars — 58
- Our solar system — 58
  - *Characteristics of the planets* — *58*
  - *The asteroids* — *61*
  - *Other planet characteristics* — *61*
  - *Major satellites of the planets* — *63*
  - *The comet belts* — *64*
- Creation of our solar system — 64
  - *Alfven theory of plasma electric currents* — *66*
- Possibility of life outside earth in our solar system — 67
- Enormous distances to the stars — 68
  - *Limitations set by the speed of light* — *70*
  - *Distribution of star sizes* — *71*
- Measurement of stellar distances — 72
  - *Parallax method for measuring the distance to a star* — *72*
  - *Classifying stars by spectra to estimate distances* — *73*
  - *Radiation spectrum from an ideal blackbody* — *74*
  - *Cepheid variable stars* — *77*
- Life cycles of the sun and stars — 77
  - *The white dwarf star* — *78*
  - *The supernova* — *79*
- The structure of the atom — 80
  - *Density of the white dwarf star* — *82*
  - *Density of the neutron star* — *83*
  - *The pulsar* — *84*

5. Hubble expansion of the universe — 85
- *What is a nebula?* — *85*
- *Measuring the radial velocity of a star* — *86*

Hubble's discovery of the universe expansion ... 87
   Meaning of the Hubble expansion ... *88*
   Modern measurements of the Hubble constant ... *89*
   The observable universe ... *89*
   The apparent age of the universe ... *90*
Theories to explain the Hubble expansion ... 90
   The Hubble expansion is Apparent ... *90*
   The Steady State theory ... *91*
   The Big Bang theory ... *92*
The Big Bang age dilemma ... 93
The Gamow Big Bang theory ... 94
The Modern Big Bang theory ... 96
Cosmic microwave background radiation ... 98

6. The Einstein theory of relativity — 101
Basis for the singularity concept ... 101
The nature of light ... 101
   What is a light wave? ... *101*
Discovery of the radio wave ... 104
The aether concept ... 105
Measuring the speed of light ... 105
   The Einstein relativity principle ... *107*
The Einstein Special theory of Relativity ... 108
   Equipment for measuring the speed of light ... *108*
   Application of relativity to a fictitious space-travel experiment ... *109*
   Symmetry of relation between observers A and B ... *112*
   Implications of relativistic effects ... *112*
The Einstein General theory of Relativity ... 114
   Generalizing the relativity principle ... *114*
   Equivalence between acceleration and gravity ... *115*
   Redshift produced by gravity ... *116*
   Effect of gravity on a time measurement ... *117*
   Effect of gravity on a distance measurement ... *118*
   Effect of gravity on the speed of light ... *118*
   The mathematical theory of curved space ... *118*
   Verification of general relativity ... *119*
   Reduction of speed of light, clock rate, and spatial dimension produced by gravity ... *121*
   The Schwartzschild limit ... *122*
   Computer solutions of general relativity ... *123*

7. Einstein's rejection of singularities — 124
The black hole singularity ... 124
   Theory of the neutron star ... *124*
   The black hole concept ... *126*
   Physical evidence for black holes ... *127*
The Big Bang singularity ... 127
Are the Einstein Equations correct? ... 129
The Yilmaz refinement of the Einstein theory ... 129

8. The Yilmaz theory of gravity — 131
Basis for the Yilmaz theory ... 131
   Exact effect of gravity on a time measurement ... *131*

Exact effect of gravity on a distance measurement	*131*
Yilmaz gravitational field equation	*132*
Time-varying solution to the Yilmaz theory	*132*
Lack of singularities in the Yilmaz theory	**132**
The black hole singularity	*132*
The Big Bang singularity	*134*
Consistency with quantum mechanics	**135**
String theory	*135*
The Yilmaz alternative to string theory	*135*
Variation of speed of light with direction	**136**
Uniqueness of the Yilmaz theory	**136**
Inter-stellar space travel	**137**

9. Lack of objectivity in astronomy — 138

The bandwagon mentality in astronomical research	**138**
The editorial of Geoffrey Burbidge	**139**
Halton Arp's quasar discoveries	**140**
The discovery of the quasar	*140*
Quasar observations of Halton Arp	*141*
What is a quasar?	*143*
Eric Lerner and Nobel laureate Hannes Alfven	**144**
The cosmology theory of Hannes Alfven	*144*
Mythological philosophy of Big Bang research	*145*

10. The Yilmaz cosmology theory — 148

Assumptions of the Yilmaz cosmology theory	**148**
Compression of the universe	**150**
Uniqueness of Yilmaz theory predictions	**152**
Creation of matter	**154**
Constancy of the universe size	**154**
Source of the continual creation of matter	**156**
Apparent variation of galaxy velocity with distance	**157**
Cosmic microwave background radiation	**158**
How can gravity make the universe expand?	**159**

11. Evidence of the mystery of Creation — 160

Concepts of Creation supported by strong evidence	**160**
Creation of life on earth	*160*
The creation of our sun and our solar system	*161*
The creation of the stars	*162*
How were the clouds of hydrogen created?	**163**
The basic theories of universe creation	*163*
Theory that the universe expansion is Apparent	*163*
Our two basic theories of creation of matter	*164*
The Big Bang theory	**164**
The Gamow Big Bang theory	*164*
Cosmic microwave background radiation	*165*
The modern (singularity) Big Bang theory	*165*
Einstein's rejection of the singularity	*166*
The Big Bang age dilemma	*168*
The Steady State theory	**168**
The Hoyle Steady State theory	*168*

The Yilmaz Steady State theory   *168*
We do not know the answer   169
What should the reader believe about Creation?   170

APPENDICES   171
A. The DNA genetic code   171
B. The nature of matter   174
    What is a molecule?   174
    Elementary particles that form an atom   174
    Chemical reactions among atoms   175
    Radiation spectrum of an atom   176
    Experiments with radioactivity   177
    Particle accelerators   178
    Nuclear fission   179
    Nuclear fusion   180
    Energy released by nuclear fusion and fission   181
    Implications of our knowledge of the atom   182
C. Density and matter in the universe   183
    C.1 Luminosity density of the universe   183
    C.2 Dark matter   183
    C.3 The source of dark matter   184
    C.4 Total mass in the universe   185
    C.5 Mass density of the universe   186
    C.6 Predicted density of matter   186
    C.7 Theoretical mass of the universe   187
    C.8 Rate of creation of matter   188
    C.9 Size of the initial Big Bang universe   188
D. Cosmic microwave background radiation   189
E. Theoretical Discussion of Einstein and Yilmaz theories   193
    E.1 The basis for the Yilmaz theory   193
       E.1.1 Einstein formula for gravitational redshift   *193*
       E.1.2 Yilmaz formula for gravitational redshift   *194*
    E.2 The Einstein gravitational field equation   195
       E.2.1 Tensors of the Einstein theory   *195*
       E.2.2 Solving the Einstein equations   *197*
       E.2.3 Limited to single-body solution   *199*
    E.3 Yilmaz gravitational field equation   **200**
       E.3.1 Tensor for gravitational energy and force   *200*
       E.3.2 Need for a gravitational field stress-energy tensor   *201*
       E.3.3 Variation of speed of light with direction   *203*
Glossary   204
Bibliography   208
Index   213-219

Preface

The Facts of Creation

How were the heavens and the earth created? How was life on earth created? How were we humans created? This book examines scientific evidence concerning Creation to help readers decide what to believe about these age-old mysteries. This includes a simple physical explanation of Einstein's general theory of relativity, which is the foundation for the Big Bang theory of the creation of our universe.

Beliefs Concerning Creation

A recent television program of the *Public Broadcasting System* discussed the issue of Creation with students and faculty of a conservative Christian college. Only two points of view were expressed. Some believed literally in the Biblical Adam and Eve Creation story, while the others completely accepted the scientific explanations. The scientific explanations included Darwin's theory of Evolution, theories of the creation of the earth and the sun, and the Big Bang theory of the creation of the universe.

In other words, the individuals either completely accepted the claims of scientists concerning Creation, or completely rejected them.

But the issues are not that simple. The evidence supporting the scientific explanations of Creation varies enormously among the various theories. In order to draw meaningful conclusions concerning the many aspects of the mystery of Creation, one should understand the evidence supporting each aspect, and decide for oneself what to believe, what to question, and what to reject, and when to conclude, "We do not know the answer".

Darwin's theory of Evolution has been hotly debated since it was presented 150 years ago. Although the evidence supporting Evolution has grown steadily over the years, opposition to it is still strong. A new aspect of the controversy has recently emerged with the research on DNA, the genetic code carried in our cells that establishes the inherited characteristics of all living organisms. The DNA evidence provides a much firmer basis for evaluating the validity of the Darwin theory.

If we accept Darwin's theory, does this mean that humans are fundamentally no different from animals? A clue to this question occurred 40 thousand years ago. Anatomically modern humans have existed for at least 100 thousand years, but it was not until 40 thousand years ago that the psychological character that we associate with human behavior became apparent in the fossil record. Prior to that time, stone implements were all utilitarian in nature. Then there suddenly appeared an abundance of decorative and artistic items in stone and bone, including beads, pendants, and the like. The aesthetic character that is uniquely associated with humanity did not evolve gradually, as evolutionary theory would predict. It emerged suddenly. Could this sudden emergence of human awareness signify Divine intervention?

Although human behavior changed drastically and abruptly 40 thousand years ago, there was no detectable physical change at that time in the human anatomy. Humans continued their nomadic hunter-gatherer life until agriculture was developed 12 thousand years ago.

Age of the Earth

The Bible has often been used to show that our earth is only a few thousand years old. However there is overwhelming evidence that the earth has existed for billions of years.

The measurement of radioactive elements in rocks has given scientists accurate means of measuring the ages of many rocks. The oldest earth rocks solidified from molten lava 4.3 billion years ago. Microscopic life appeared at least 3.6 billion years ago. The first animals appeared 600 million years ago, and the first vertebrate (a primitive fish) appeared 500 million years ago. Vertebrates moved onto the land (as amphibians like frogs) 360 million years ago. Gigantic dinosaurs controlled the land for 120 million years, until they were eliminated by a catastrophe 65 million years ago. At that time our modern Age of Mammals began.

Creation of the Stars

Our knowledge of nuclear physics, derived from developing atomic nuclear bombs and nuclear power plants, has given scientists detailed understanding of the nuclear reactions occurring within the sun and stars, which generate their enormous energy.

By studying the light received from stars, astronomers have derived extensive knowledge of the characteristics of stars within our galaxy. By combining this with the knowledge of nuclear processes inside the stars, reliable theories have been developed showing how our sun and the stars were created, how they live, and how they will eventually die.

There is strong evidence that our sun (like other stars) was created when an enormous cloud, mostly of hydrogen gas, coalesced because of gravity. When our sun was formed, our solar system (including the earth) was created. The creation of a solar system appears to be a normal aspect of stellar development, and so we should expect that many stars in our Milky Way galaxy may have planets similar to the earth.

Creation of the Universe

Our Milky Way galaxy contains 100 billion stars. As astronomers look outside our galaxy, they see billions of similar galaxies that extend to the limits of their telescopes. These galaxies are moving away from us at velocities proportional to distance. Our universe is expanding. Galaxies are flying apart as if our whole universe has emerged from an enormous *Big Bang* explosion.

Nearly all astronomers accept as established fact the concept that our universe began about 15 billion years ago as an extremely dense mass that exploded with a *"Big Bang"*. Nevertheless, when scientists attempt to extrapolate the universe backward to the instant of the *Big Bang*, they encounter serious contradictions.

The Gamow Big Bang Theory

The father of the Big Bang theory was George Gamow, a nuclear physicist who was a leader in the development of the atomic nuclear bomb during World War II. In 1947 he presented his Big Bang theory, postulating that the universe initially had the density of matter in the atomic nucleus, which weighs one billion tons per teaspoon. Gamow considered nuclear matter to have the greatest density that is physically possible.

Assuming that the universe is expanding uniformly, measurements of the universe expansion predict that a galaxy 15 billion light years away should be moving at the speed of light, and so should not be observable. Hence our observable universe is considered to be a sphere with a radius of 15 billion light years.

Astronomical data indicate that the stars shining within our observable universe have a mass equivalent to 60 billion times one billion suns. Galaxy motions show that there must be about 300 times as much dark matter (which we cannot see) as there is luminous matter. If we assume that all of this matter (shining stars plus dark matter) within our observable universe was originally compressed into a sphere with the density of nuclear matter (one billion tons per teaspoon), the initial universe would have just about fit within the orbit of the planet Mars.

The Modern Big Bang Theory

The Gamow Big Bang theory was widely supported for about two decades, when it was replaced by the Modern Big Bang theory, which is based on the concept of the *singularity*, in which matter is ideally compressed to infinite density.

James Peebles was acclaimed to be the "father of modern cosmology" by the *Scientific American* (Jan. 2001, p. 37). In the Oct. 1994 *Scientific American* (p. 53), Peebles concluded that at the instant of the Big Bang the observable universe was "smaller than a dime".

On the other hand, most Big Bang cosmologists now endorse the *"inflation"* postulate of universe creation, which concludes that the observable universe was initially "one trillionth of the size of a proton". (A proton is 2 trillionths of a millimeter in diameter.)

The Einstein General Theory of Relativity

Why have astronomers discarded the Gamow Big Bang postulate and replaced it with the singularity? Why have they squeezed the observable universe from the size of the orbit of Mars to the "size of a dime", and probably to "one trillionth of the size of a proton"? Yet the renowned nuclear physicist George Gamow considered his postulate to represent the greatest possible density of matter.

The answer is that modern Big Bang theories are derived from computer studies of the Einstein general theory of relativity, which were first performed about a decade after Einstein's death in 1955. These computer studies have predicted that our universe began as a singularity. **Nevertheless, throughout his lifetime, Einstein absolutely rejected all singularity predictions derived from his theory.** In 1945, Einstein recognized that his theory implied a singularity at the beginning of the universe expansion. He insisted that his equations would only apply approximately under such extreme density of matter, and so cannot be used to predict a singularity.

Bandwagon Philosophy in Astronomy

With modern instruments, improved telescopes, and satellite sensors, astronomical knowledge has advanced greatly in recent years. However, economic forces have produced a bandwagon mentality that is choking astronomical research. Dissent is suppressed, and without open scientific debate there can be no true science.

The case of the eminent astronomer Halton Arp proves this point. He received his PhD degree in astronomy in 1953 and performed 30 years of distinguished research at the Palomar and Mt. Wilson Observatories, receiving many awards for his findings. Starting in the mid 1960's, he began extensive observations of quasars, which led him into conflict with accepted astronomical dogma.

A quasar superficially looks like a star, but has a very large spectral shift, which seems to indicate that it is moving away from us at an enormous speed. With such a speed, it would have to be billions of light years away, and at such a great distance it would have to radiate an unbelievable amount of energy for us to see it. The explanation of a quasar would be much more reasonable if something other than velocity is causing its large spectral shift, and that quasars are much closer than most astronomers believe.

Halton Arp performed many astronomical observations of quasars, which provided strong evidence that quasars are relatively close. Those observations include filament structures that directly connect quasars to galaxies having much smaller spectral shifts.

The astronomical community strongly objected to Arp's research, because it contradicted the widely accepted concept of the quasar. In 1984, the committee that controls observation time at the Palomar and Mt. Wilson Observatories refused to allow Arp to use these facilities. After 30 years of distinguished astronomical research, Arp was forced to accept early retirement, and moved to the Max Planck Institute in Munich, Germany to continue his career.

Chapter 9 presents the case of Halton Arp, along with other examples of the dogmatic approach of astronomers. For example, Hannes Alfven received the Nobel Prize for his pioneering research on plasma physics. Nevertheless, astronomers reject papers on plasma physics despite their revolutionary implications in astronomy

Because the astronomical community is suppressing facts that do not agree with its accepted concepts, it is incapable of deriving objective conclusions from its observations. Astronomers cannot determine which of their conclusions are based on solid evidence, and which are due to false interpretations.

Understanding the Einstein Theory

Many concepts of astronomy have been derived from the Einstein general theory of relativity, including the modern *Big Bang* theory and the *black hole*. This book explains the Einstein theory, to give readers the insight to evaluate the claims of astronomers concerning Creation.

Although the mathematical equations of general relativity are very complicated, this book shows that the physical principles of the Einstein theory can be explained in a manner that is readily understood by the average reader.

The Yilmaz Theory of Gravity

In 1958 Huseyin Yilmaz published his gravitational theory in the prestigious *Physical Review*. The Yilmaz theory satisfies the principles of general relativity, and so is a refinement of the Einstein theory.

The Einstein theory is specified by a very complicated formula called the Einstein gravitational field equation, which represents 10 independent equations. Einstein obtained this formula in an intuitive manner, after years of searching for an answer. This equation appeared to work, and so Einstein accepted it.

While performing PhD research on general relativity at the Massachusetts Institute of Technology, Yilmaz discovered the key that allowed him to derive a different gravitational field equation by rigorous analysis. The Yilmaz theory does not allow the singularity predictions that have been obtained from the Einstein theory. The Yilmaz theory is very much easier to apply than the Einstein theory, because Yilmaz has a general solution to the formula that specifies his theory.

After powerful computers became widely available in the 1960's, an enormous amount of research has been performed to achieve computer solutions of the complicated Einstein equations. If the Yilmaz theory were accepted, the value of this research would be degraded.

When the Yilmaz theory is applied to cosmology, it yields a *"Steady State"* explanation of the universe, which predicts that the universe is infinitely old and that diffuse matter is continuously created throughout the universe to compensate for the universe expansion. The theory suggests that this diffuse matter is derived from energy radiated by dying stars.

The Yilmaz theory predicts that the universe expansion is a natural relativistic effect caused by gravity. It predicts that the relativistic effects of gravity distort space, so that dimensions at great distances are strongly compressed as seen by an observer on earth. Because of this relativistic compression, the over-all size of the universe should remain constant, even though the universe expands locally about every point.

We Do Not Know the Answer

Astronomers insist on treating the Big Bang theory as fact, even though it is riddled with contradictions. A consequence of this philosophy is that observational evidence and theoretical studies that are inconsistent with the Big Bang theory are rejected as being irresponsible. This serious problem was eloquently discussed in a 1992 essay of the *Scientific American* by the eminent astrophysicist, Professor Geoffrey Burbidge, which is summarized in Chapter 9.

The singularity represents an essentially infinite density of matter, and so grossly conflicts with physical evidence. Nevertheless, astronomers insist that black hole singularities physically exist, and that our universe began as a Big Bang singularity. *The only evidence supporting the physically unbelievable singularity is the Einstein general theory of relativity; yet, Einstein absolutely rejected all singularity predictions derived from his theory.* As shown on page 128, Einstein recognized in 1945 that his theory would not apply accurately under conditions of extreme density of matter, and so cannot be used to predict a singularity.

This book describes the Yilmaz theory of gravity, which applies the principles of general relativity, and so is a refinement of the Einstein theory. Since the Yilmaz theory does not allow a singularity, it demonstrates, in agreement with Einstein, that "singularities do not exist in physical reality" and are inconsistent with relativity principles.

The Yilmaz Steady State cosmology theory, which is derived from the Yilmaz theory of gravity, shows that there is at least one viable explanation of the universe that does not include the Big Bang explosion. Whether or not one endorses the Yilmaz theory, it demonstrates that the validity of the Big Bang concept has definitely not been proven. Astronomers are severely violating scientific principles when they treat the Big Bang theory as fact, rather than theory.

How was our universe created? The scientific response should be, **"We do not know the answer"**.

xx   Beliefs and Facts of Creation

Figure 1-1: The M51 Whirlpool galaxy, which is 35 million light years away, resembles our own Milky Way galaxy. Its smaller companion galaxy is NGC5195.

Chapter 1

Introduction

Our Search for the Truth

Since the dawn of human awareness, we have wondered, "How were we created? How were our earth and heavens created?"

All societies have traditional explanations that attempt to answer these eternal mysteries. Jews and Christians looked to the Bible, which gives the following story of Creation:

> *"In the beginning, God created the heavens and the earth. And the earth was without form and void, and darkness was upon the face of the deep. And the spirit of God moved upon the face of the waters. And God said, 'Let there be light', and there was light."*

Fanciful as this explanation may seem, we will see that it is quite consistent with scientific evidence.

We shall examine the scientific evidence relating to these great mysteries. As we search for truth, we will find that we must treat many scientific proclamations with skepticism. This book will separate those scientific concepts that are strongly supported by evidence from those that rest on shaky ground. This should allow the readers to use their own reasoning to find the truth.

The confidence with which a group of scientists proclaims a concept can have little relation to the degree of evidence supporting that concept. Strong debate is a sign of healthy inquiry, and is essential in the search for scientific knowledge. Uniformity of thought can represent a bandwagon mentality that is covering up great gaps in the evidence. A major goal of this book is to determine the different levels of scientific evidence behind the various concepts, and thereby give the reader a firm basis for deciding what to believe.

Along with scientific evidence, people often use their intuition as a guide for finding eternal truths. We feel the overwhelming beauty of the natural world that surrounds us. --- in the radiant flowers, in the majestic forests, in the inspiring waters of a lake, in the birds with their delightful colors, in the ever-changing clouds against a deep blue sky. To me, the most inspiring sight is my view of the stars in the night sky.

Out Mystical Milky Way

Nearly every summer I visit a rural location that is far from city lights. I am always excited to view the sky on a clear moonless night, to see countless stars shining in wondrous majesty. The most inspiring feature to me is that pale white pathway across the sky, called the Milky Way. The Milky Way encircles the celestial sphere, dividing it into nearly equal hemispheres. It is sad that many people today have never seen our mystical Milky Way, because it is obscured by sky glow due to the reflection of electric lights from the atmosphere.

The Milky Way is our view of 100 billion stars that form our Milky Way galaxy, which is 100 thousand light-years in diameter. A light-year is the distance that light travels in one year, which is about 10 trillion kilometers. Our galaxy is similar to the Whirlpool galaxy in Fig. 1-1 and on the front cover. Our sun is 2/3 of the distance from the center to the circumference and lies within a spiral arm. (The Whirlpool galaxy is 35 million light-years away, and is 2 degrees from the end of the Big Dipper handle.)

Prior to 1900 it was generally believed that our Milky Way galaxy was the whole universe. Then astronomers developed means of measuring stellar distances and discovered that many of the fuzzy astronomical objects called "nebulae" were actually distant galaxies containing billions of stars like our own Milky Way. Our Milky Way galaxy is merely one out of billions of galaxies that comprise our universe. The size of our universe seems beyond comprehension.

Yet there is something even more unbelievable about our universe. The universe appears to be flying apart. Galaxies are moving away from us at velocities proportional to distance. What can this mean? What does it tell us about how our universe was created?

But before we consider the great mystery of our universe, let us begin with the simpler issues: the creation of our sun and our earth.

Creation of Our Sun and Our Earth

Nuclear Fusion and Nuclear Fission

The enormous energy generated by our sun is produced by nuclear fusion occurring at the center of the sun. Atoms of hydrogen (the lightest element) are fused to form helium (the next lightest element). A small reduction of mass occurs, and this loss of mass is converted into a tremendous amount of energy, which is radiated by our sun.

With the development of the atomic nuclear bomb during World War II, physicists learned a great deal about atomic nuclear energy. This has provided a firm basis for analyzing the nuclear processes that occur within our sun and the stars.

Nuclear theory predicts that the temperature at the center of the sun is 15 million degrees Celsius. The pressure is so great that the hydrogen at that point is compressed to 200 times the density of water. At this extreme temperature and pressure, the nuclei of the hydrogen atoms are forced nearly into contact with one another. This causes four atoms of hydrogen to fuse together and form one atom of helium.

The helium atom has 0.710 percent less mass than the four hydrogen atoms that form it, and this loss of mass is converted into enormous energy. If one gram of matter were converted completely into energy, it would produce 25.0 million kilowatt-hours of energy. Taking 0.71 percent of 25 million kilowatt-hours gives 177,500 kilowatt-hours. Therefore the fusion of one gram of hydrogen to form helium releases 177,500 kilowatt-hours of energy. One gram is one third of the weight of a United States penny (one cent coin).

The mass of our sun is 2000 trillion times one trillion metric tons. (One metric ton is 1000 kilograms.) The sun converts about 600 million metric tons of hydrogen into helium every second. It releases about 100 billion times one billion kilowatt-hours of energy every second, which is a power of 360 billion times one trillion kilowatts.

The atomic nuclear bombs dropped on Japan during World War II used nuclear *fission* (not nuclear *fusion*). Atomic nuclear power plants also use nuclear *fission*. In nuclear *fission*, very heavy atoms are split almost in two in a chain reaction to form lighter atoms. The nuclear processes implemented in the stars use nuclear *fusion* in which light atoms are fused to form heavier atoms. Nuclear *fusion* is implemented on earth in the hydrogen bomb, which was first tested in the Eniwetok Atoll in November 1952. A hydrogen bomb using nuclear *fusion* is much more powerful than the nuclear *fission* bombs that were dropped on Japan in August 1945 to end World War II.

In a hydrogen nuclear fusion bomb, a nuclear fission bomb is exploded into a body of hydrogen. The explosion of the nuclear fission bomb creates the enormous temperature and pressure required to achieve nuclear fusion, thereby producing the much more powerful nuclear fusion explosion.

The word "fission" was originally used to describe the process whereby a bacteria cell splits in two to reproduce. In 1939, it was discovered that atoms of the very heavy element uranium can split in a chain reaction, and thereby release tremendous energy. Since the uranium atom splits almost in two, the process was termed "nuclear fission".

Further explanations of nuclear fusion and nuclear fission are given in Appendix B. This discussion of nuclear energy shows that physicists have developed detailed theoretical and experimental knowledge of nuclear energy. Consequently, they can compute with high confidence the nuclear processes being implemented within the sun and the stars.

Formation of Our Sun and the Stars

The processes involved in the creation of a star take far too long to see the changes that they produce. However, astronomers have been able to learn about these changes by observing many stars of different ages. These astronomical observations have been combined with analyses of nuclear energy to achieve a reliable picture of the major steps of stellar development. This has yielded the following story.

A star begins as a diffuse cloud that is mostly hydrogen gas, but also contains a small amount of dust particles that make up the solid matter from which solid planets like the earth are formed. The gas cloud collapses because of gravitational attraction among the gas molecules. As the cloud collapses, gravitational energy is released, because the gas molecules are falling, and this energy heats the cloud. When the center of the cloud reaches 15 million degrees Celsius, nuclear fusion is ignited and a star begins to shine. The star can shine for billions of years as the nuclear fusion process converts the hydrogen into helium.

The size of the cloud that forms a particular star determines the mass of the star. The larger the star, the brighter it shines, and so the faster it burns up its fuel. Stars range in mass from about 1/10 of the sun's mass to 100 times the sun's mass. If a star is formed with less than 1/12 of the sun's mass, it does not have enough mass to achieve nuclear fusion, and so it never shines as a true star.

For stars near the size of our sun, the power radiated is proportional to the fourth power of the mass. Thus a star with twice the mass of our sun radiates 16 times the power of our sun; a star with half the mass of our sun radiates 1/16 of our sun's power.

Our sun has been fusing hydrogen to form helium for 5 billion years, and will continue to do this for another 5 billion years, until the hydrogen nuclear fuel is exhausted. Then the sun will enter the twilight period of its life, which is explained in Chapter 4.

The more power a star radiates, the shorter is its life. A star with twice the mass of our sun has 1/8 of the sun's life span. A star with half the mass of our sun has 8 times the sun's life span.

Formation of Our Solar System

Our planets were created in the following manner. When our sun was formed, a disk of gas and dust rotated around the sun, somewhat like the rings that rotate around the planet Saturn. The gas and dust in this disk gathered into larger and larger bodies, until a limited number of large bodies remained (the planets), each at greatly different distances from the sun. If two bodies were reasonably close, they would eventually collide, and the larger body would swallow the smaller one.

As will be shown in Chapter 4, the formation of a disk of gas and dust around a star appears to be a normal occurrence in the creation of a star. Consequently we should expect that many stars within our Milky Way galaxy have solar systems, and many of these may have planets like earth that can support life. However, most of these planets are at vast distances, so far we may never be able to reach them.

There are nine planets in our solar system. The four inner planets, Mercury, Venus, Earth, and Mars, are solid and relatively small. The next four, Jupiter, Saturn, Uranus, and Neptune, are gaseous giants. The outermost planet, Pluto, is solid and small, about the size of our moon. Pluto is unlike the other planets, and may be a captured comet.

Even the gaseous planets have large solid cores, which suggests that all of the planets began as solid bodies. When the solid core of a giant planet became sufficiently massive to hold light gasses, it gathered up the hydrogen and helium, and grew rapidly. The pressure of light from the sun forced the gasses away from the sun, and so there was little hydrogen and helium in the region of the inner planets. This reasoning explains why the inner planets are small and solid, and the outer planets are much larger, and have low density. The least dense of the outer planets is Saturn, which is so light it could float on water.

Creation of the Earth

The earth and the other three inner planets were formed from dust in the disk that rotated around the sun. The dust particles congregated into larger and larger bodies. When a small body collided with a large body, it released tremendous energy, which kept the large body hot and molten. Hence the earth was formed as a molten sphere.

The earth reached essentially its full size about 4.6 billion years ago. The small bodies that continued to collide with the earth after the earth was formed are called meteorites. The molten rock that originally covered the earth was like the lava that is now released by volcanoes.

Scientists can measure the age of molten (volcanic) rock since the lava hardened, by examining radioactive elements in the rock. By measuring the amount of the original radioactive element, and the amount of the elements produced by radioactive decay, they can determine the fraction of the radioactive element that has decayed since the rock hardened. Scientists know the rate of radioactive decay for an element, and so can determine the time that has elapsed since the rock hardened.

When the rock was molten, the elements produced by radioactive decay moved away from the radioactive element, and so are not present next to the radioactive element in the hardened rock.

The oldest earth rocks are about 4.3 billion years old. This suggests that the crust of the molten earth hardened about 4.3 billion years ago. A meteorite fell in New Mexico that is 4.5 billion years old. This gives another basis for estimating the age of the solar system.

We have discovered a great deal about the early stage of earth development in our study of rocks obtained from the moon. The craters on the moon were caused primarily by meteorite collisions. Since the moon and earth have received similar meteorite impacts, we can learn about early meteorite collisions on the earth by examining the moon. Early meteorite craters on the earth have been erased by weather and volcanic activity. There is no weather on the moon, and volcanic activity has been limited.

Study of moon rocks have shown that the earth was heavily bombarded by meteorites for 500 million years after the earth crust hardened 4.3 billion years ago. Thus the earth had a hellish environment until 3.8 billion years ago.

The excerpt from the Bible that began this chapter is consistent with our scientific discussion of Creation. A cloud of hydrogen collapsed until it reached sufficient temperature to ignite nuclear fusion, and our sun began to shine. The Bible tells us, *"And God said, "Let there be light', and there was light."* Our earth began as a collection of dust particles rotating around the sun. As the Bible says, *"In the beginning, the earth was without form and void, and darkness was upon the face of the deep".*

When the earth was molten and very hot, much of the water on earth may have boiled away. Meteorites contain a large amount of water. Hence the water that fills our oceans may have come largely from meteorites that hit the earth after its crust hardened.

Water has been essential in the creation of life. Life on earth began in our oceans.

Creation of Life on Earth

There is evidence that very simple microbes called archaea lived at least 3.8 billion years ago. (Archaea are discussed in Chapter 2). The first clear evidence of life is demonstrated by structures called stromotolites that are 3.6 billion years old and were colonies of specialized bacteria called cyanobacteria.

Cyanobacteria contain the chemical chlorophyll, which implements photosynthesis, the process used by plants to derive energy from sunlight. In photosynthesis, chlorophyll uses the energy from sunlight to synthesize carbohydrate food, by combining the carbon in carbon dioxide with the hydrogen in water.

For about 2 billion years, life on earth consisted only of microscopic single-celled organisms. Then 1.8 billion years ago, the first multi-celled organism appeared. This was seaweed, a marine plant.

The first multi-celled animals appeared 600 million years ago, as soft-bodied creatures like jellyfish. Then, 545 million years ago, an explosion of animal life occurred in our oceans, at the start of the Cambrian geological period. Within 10 million years, representatives from nearly all of the animal phyla suddenly appeared. This included a primitive *chordate*, the phylum that includes vertebrates, which are animals with backbones. (Phyla are the basic categories for classifying animal types.)

The first vertebrate appeared 510 million years ago. This was a primitive jawless fish, like the lamprey. Advanced fishes evolved, and from these came the amphibians, which moved onto the land 360 million years ago. Amphibians (which include frogs) lay small fragile eggs in water, like fish, and begin life like fish as tadpoles. The first reptile appeared 335 million years ago. Reptiles lay amniotic eggs, like bird eggs, that can hatch on land.

Reptiles reached their zenith with the dinosaurs, which dominated the world between 225 and 65 million years ago. The dinosaurs were suddenly driven into extinction by a worldwide catastrophe, probably caused by the impact from a large meteorite about 10 kilometers in diameter. With the dinosaurs eliminated, mammals took control of the land, which they have dominated for the last 65 million years.

The first ape appeared 25 million years ago. About 6 million years ago, an African ape, called Australopithecus, developed the ability to walk upright. It had the brain of a chimpanzee, but could walk upright like a human.

The first stone tools appeared 2.5 million years ago, apparently made by Homo Habilis, which walked upright and had an appreciably larger brain then Australopithecus. Homo Erectus appeared 1.8 million years ago. He had a larger brain than Homo Habilis, and made more sophisticated stone tools. Homo Erectus spread from Africa to Europe and Asia.

Anatomically modern humans appeared at least 100 thousand years ago, and may have existed 50 thousand years earlier. About 40 thousand years ago, the fossil record of humans suddenly began to display a radical cultural change, showing many articles of stone and bone of an artistic character.

Humans continued their nomadic hunter-gather way of life, until agriculture was developed 12 thousand years ago. Since agriculture allowed people to live in large communities, towns and cities evolved, and our civilizations began.

Darwin's Theory of Evolution

The fossil record shows a progression of life on earth, which began as microscopic organisms, nearly 4 billion years ago, and gradually led to the complex world that we experience today. To explain this progression of life, Charles Darwin presented his theory of Evolution in 1857. This theory was highly controversial when it was presented, and many people today still dispute it.

Principle of Natural Selection

Darwin based his theory on the principle of selective breeding, which has radically changed the characteristics of plants and animals since agriculture began, 12 thousand years ago. In each generation, one selects for breeding the specimens that have the most desirable characteristics. By repeatedly choosing those specimens with the best characteristics of each generation, a plant or animal is developed that is radically different from the original stock.

Domesticated cattle are much smaller (and more easily handled) than the wild aurochs from which they were derived. Corn has changed so much it can no longer reproduce in the wild. The many breeds of dogs and horses are the result of selective breeding over thousands of years.

Darwin concluded that selective breeding exists throughout the world as a natural process. Random changes of the genetic characteristics of organisms are continually occurring. Many of these genetic changes produce undesirable characteristics. The individuals who inherit them fare poorly in the competition for life, and have a low probability of reproducing. Some genetic changes yield desirable characteristics, and so the individuals who inherit them reproduce prolifically. By this "natural selection" process, new breeds and new species are continually being developed, to produce organisms that are best suited to the environment. Darwin called this process, "survival of the fittest".

By means of "natural selection" and "survival of the fittest", Darwin postulated that organisms have gradually evolved over many millions of years to produce the enormous variety of life that we observe today.

Objections to the Darwin Theory

Many people have strongly objected to the Darwin Evolution theory on religious grounds. An obvious implication of the theory is that humans and apes have a common ancestor; we are distant cousins of the apes. Carrying the reasoning further suggests that we are related to all vertebrates, and maybe to all living organisms.

Opponents of evolution have argued that the fossil record is very limited. It is impossible to prove from the sketchy fossil record that humans are directly related to the animals. The Darwin theory of Evolution is merely a theory, and will always remain a theory that cannot be proven.

To support his theory, Darwin used evidence concerning the characteristics of living organisms. The great similarity of the physiology of closely allied species must mean that they are biologically related. They must have had a common ancestor.

Medical science has greatly advanced by performing experiments on animals. How could these experiments on animals have applicability to humans unless animals and humans are biologically related?

Yet the opponents of Evolution continually maintain that none of this evidence is conclusive. They insist that the theory of Evolution is merely a theory, not a scientific principle.

The DNA Evidence

Over the past half century our knowledge of the process of inheritance has advanced tremendously with the studies of DNA. On April 14, 2003, scientists reported that they had deciphered the complete DNA code for a human, which consists of 3 billion basic components, called nucleotides. There are four different nucleotides, which are usually identified by their initial letters, A, C, G, and T. The 3-billion long arrangement of these four basic components specifies the complete inheritance characteristics of a human being. Further discussion of DNA is given in Appendix A.

The DNA code is not just a theoretical concept. It is a powerful scientific tool. DNA evidence is used to solve crimes and to determine who is the father of a child. DNA techniques have been used to produce new organisms. For example, the gene from a bacterium was inserted into the DNA of a plant to produce a plant that can manufacture vitamin A, thereby greatly aiding the nutrition of impoverished populations.

Studies have shown that the human DNA sequence of 3-billion elements closely matches the DNA sequence of the chimpanzee. Early studies concluded that the match was greater than 98 %, but a recent study has found that the match is 95 %. [63] The early studies only considered substitutions of nucleotides in the 3-billion element DNA code, but the new study also included insertions and deletions.

It has been argued that this DNA similarity between humans and chimpanzees is misleading. After all, the chimpanzee has 24 pairs of chromosomes, whereas the human has 23 pairs. Although the full human DNA genome sequence has been determined, we know only part of the chimpanzee DNA sequence.

Despite these objections, the DNA similarity between chimpanzees and humans is enormous, no matter how one interprets the data. How is it possible to have such similarity in the 3-billion DNA components of humans and chimpanzees? In the author's opinion, the only possible answer is that humans and chimpanzees must have had a common ancestor. It does not seem possible that this similarity could have resulted from chance. The readers should decide for themselves what to believe about this controversy.

The Emergence of Human Awareness

Since DNA evidence apparently demonstrates that humans and chimpanzees are biologically related, it seems to verify the basic principle of the Darwin theory of Evolution. Does this mean that humanity is nothing more than a biological accident? Not really.

Human awareness, the characteristic that fundamentally separates humans from animals, emerged suddenly about 40 thousand years ago. This change has greatly perplexed anthropologists, and was far too abrupt to be explained by evolution. With the suddenness of this change, it seems legitimate to raise the question, "Was this event the result of Divine intervention?"

Anthropologists separate the stone-age culture of human evolution over the past million years into the *Lower Paleolithic* or Old Stone Age (prior to 250,000 years ago); the *Middle Paleolithic* or Middle Stone Age (from 250 to 40 thousand years ago), and the *Upper Paleolithic* or New Stone Age (from 40 to 10 thousand years ago). Agriculture began about 10,000 years ago, and quickly led to modern civilizations. There was a dramatic cultural change 40,000 years ago at the boundary between the Middle and Upper Paleolithic periods. [61] (Chapter 5)

Lewin [61] (p. 115) writes, "In the European prehistoric record - - - an abrupt transition occurred between 40,000 and 30,000 years ago. Stone tool technologies, essentially unchanged for 200,000 years, were suddenly replaced by more sophisticated, stylish, and rapidly evolving artifact traditions. Body decoration (in the form of beads, pendants, and possibly necklaces) - - - appeared for the first time."

Lewin [61] (p. 122) states, "Neanderthals and anatomically modern humans appear to have coexisted from about 100,000 years ago to perhaps 50,000 years ago. - - The puzzle, however, is that from all of the archeological evidence so far recovered - - there was no technological difference between the Neanderthals and their anatomically modern human neighbors, at least in the range of artifacts they manufactured."

The fossil record shows that anatomically modern humans have existed for 100 thousand years, and possibly for 150 thousand years. By this we mean that fossilized remains have been found that are indistinguishable from the skeletons of modern humans. Anatomically modern humans lived at the same time as the Neanderthals, which existed from about 200 thousand years ago to 35 thousand years ago.

The Neanderthal had as large a brain as a modern human, but the brain was shaped differently. The forehead sloped backward, the chin receded, there were heavy eyebrow ridges, and the nasal region of the face was pushed forward. Neanderthals had very large bones, showing that they were much stronger than modern humans.

There appears to have been limited contact between Neanderthals and anatomically modern humans, because they lived in separate regions. Neanderthals were cold adapted, and lived in the ice-age climates of northern Europe and Asia; anatomically modern humans lived in warmer climates.

Until about 40 thousand years ago, anatomically modern humans and Neanderthals made essentially the same stone tools. These tools were all utilitarian in nature. For at least 60 thousand years (from 100 thousand years ago to 40 thousand years ago) anatomically modern humans and Neanderthals lived essentially the same kind of lives, as far as can be determined from the fossil record.

Then, about 40 thousand years ago, modern humans suddenly began to make a multitude of objects in stone and bone that had an aesthetic and decorative function. This showed a radical change in culture. *Human awareness*, the characteristic that truly separates humans from animals, had suddenly appeared. Neanderthals did not share in this great cultural change, and 5 thousand years later the Neanderthals became extinct.

What caused this radical change of human behavior 40 thousand years ago? As will be shown in Chapter 2, Ian Tattersall, Curator of Anthropology at the *American Museum of Natural History* in New York, has postulated that this cultural change was produced by the development of sophisticated language, which allowed modern humans to communicate much more effectively.

However, even if we accept the Tattersall postulate, we are still left with the perplexing question, "Why did this development of language occur so suddenly?" The sudden emergence of *human awareness* 40 thousand years ago is a phenomenon that cannot be attributed to biological evolution and has not been explained by anthropologists.

How Life Evolved on Earth

Chapter 2 addresses the question, "How did life evolve on earth?", and pieces together paleontology evidence to provide a coherent story. Although details of the explanation are certainly open to question, the story should help the reader understand the evidence that paleontologists have discovered concerning the creation of life on earth.

Chapter 2

The Creation of Life on Earth

The history of our earth and the evolution of its life have been deciphered from rocks and the fossils they contain. How old is our earth? A meteorite fell in Arizona that is 4.5 billion years old, and so our solar system should be at least that old. Rocks obtained from the moon have yielded extensive information concerning the formation of the earth, because the moon and the earth have experienced similar meteorite impacts.

From studies of earth and moon rocks, scientists have concluded that the earth was formed as a molten body 4.6 billion years ago. The earth had been kept in a molten state as it formed, because it was continuously bombarded by smaller bodies, which released enormous energy. The earth's crust solidified 4.4 to 4.3 billion years ago, but the earth continued to be bombarded heavily by meteorites until 3.8 billion years ago.

Early Microscopic Life

For many years, scientists considered the first definite sign of life to be *stromotolites*, which are fossils of colonies of cyanobacteria. These were created 3.6 billion years ago, and similar structures are formed today. Cyanobacteria contain the chemical chlorophyll, which implements photosynthesis by absorbing sunlight to create food. Photosynthesis is the process used in plants, which combines the energy from sunlight, the hydrogen from water, and the carbon from carbon dioxide to synthesize carbohydrates.

However, photosynthesis is too sophisticated to have been the first basis for nourishing life. Something must have preceded cyanobacteria.

Scientists have only recently found the key to this mystery of how life began on earth, with the discovery of volcanic vents on the ocean floor. This discovery was made on February 19, 1977 by the submersible vehicle Alvin in a 9200-foot dive off the Galapagos Islands. These vents are surrounded by extensive life feeding on microscopic cells that derive nourishment from hydrogen sulfide and other chemicals emanating from the volcanic vents. These microscopic cells were first thought to be bacteria, but DNA studies have shown them to be radically different. They have been given the name *archaea*. Archaea are similar to bacteria but can tolerate very harsh environments, including very high temperatures. Archaea were first clearly identified in hot springs, including those in Yellowstone National Park.

The primary building blocks of life are amino acids, which are discussed in Appendix A. The volcanic vents on the ocean floor have the energy and chemicals that could have synthesized the basic amino acids by inorganic means. Consequently, many biologists believe that archaea were the first organisms on earth. Archaea could have flourished in the hostile environment, 4.3 to 3.8 billion years ago, when the earth was heavily bombarded by meteorites.

As explained in *Science News* (May 1, 2004, pp. 280-281), there is evidence that at least some of the basic amino acids on which life is built are produced in dust clouds of space. These might have fallen to earth in meteorites before life on earth began.

The membranes of archaea are chemically different from those of bacteria. Chemical evidence has identified archaea in rocks 3.8 billion years old, but archaea may have evolved much earlier. Microscopic archaea cells probably were the only life on earth until the meteorite bombardment ended 3.8 billion years ago. Then the more delicate bacteria could survive.

The first bacteria probably fed on archaea, which derived their energy from chemicals in volcanic vents. In time, certain bacteria began to synthesize chlorophyll and thereby implement photosynthesis. Sunlight was now used as the basis for creating food that supports life. Today the energy derived from sunlight dominates life processes on earth, so much so that biologists have only recently discovered that the energy feeding some archaea comes from an entirely different source.

Archaea and bacteria are tiny cells that lack nuclei. For over one billion years, these simple cells comprised the life on earth. Then 2.7 billion years ago a new type of cell evolved, called the eukaryote. A eukaryote cell is much larger than an archaea or bacteria cell, and contains a nucleus. [45] More complicated biological processes occur within the eukaryote cell, and it allowed the evolution of multi-celled organisms. All multi-cell organisms consist of eukaryote cells. The DNA of eukaryotes is more similar to archaea than to bacteria.

Summary of the Evolution of Life on Earth

Table 2-1 lists the primary events that have been involved in the evolution of life on earth. We have just discussed events (1) to (6). For billions of years, life on earth consisted of single-celled organisms. Some of the eukaryote cells performed photosynthesis, and were called algae. Multi-celled organisms began when algae cells formed the first marine plant (seaweed), which appeared 1.8 billion years ago (7). Terrestrial plants evolved much later. (Do not confuse the ( ) referencing items of this table with the [ ] brackets that refer to the bibliography.)

Table 2-1: Major events in the evolution of life on earth

Event	Years Ago
(1) earth formed	4.6 billion
(2) earth crust hardened	4.3 billion
(3) meteorite bombardment ceased	3.8 billion
(4) evidence of archaea	3.8 billion
(5) cyanobacteria colonies	3.6 billion
(6) eukaryotes	2.7 billion
(7) seaweed	1.8 billion
(8) first animals	600 million
(9) explosion of animal life (Cambrian)	545 million
(10) first vertebrate (jawless fish)	510 million
(11) jawed fish	440 million
(12) first terrestrial plants (mosses)	430 million
(13) sharks	415 million
(14) modern bony fish	400 million
(15) ferns and related plants	400 million
(16) amphibians	360 million
(17) reptiles	335 million
(18) conifer and cycad plants	300 million
(19) dinosaurs	225 million
(20) mammals	215 million
(21) Age of Dinosaurs	205-65 million
(22) flowering plants	140 million
(23) birds	135 million
(24) Age of Mammals	65 million to present

Photosynthesis performed by cyanobacteria, single-celled algae and seaweed, released oxygen into the atmosphere as a byproduct. The oxygen finally reached sufficient concentration to support animal life. Jellyfish and other soft-body animals appeared about 600 million years ago (8). An explosion of animal life appeared in the oceans 545 million years ago, at the start of the Cambrian geological period (9). Within 10 million years, representatives from nearly all of the animal types (phyla) appeared. One of these was a primitive representative of our own phylum, the *chordates,* which includes the vertebrates (animals with backbones).

The first vertebrate (10) was a jawless fish, which appeared 510 million years ago. Fishes with jaws (11) evolved 70 million years later. Sharks (13) appeared 415 million years ago, and modern bony fish appeared 400 million years ago (14).

The first terrestrial plants (12) appeared 430 million years ago. They were low plants similar to mosses. Ferns and related plants (15) appeared 400 million years ago. The fern-like plants had vascular systems for conducting fluids, which allowed them to grow tall, some reaching 130 feet. Fern-like plants formed massive coal deposits that we now use for fuel. All of these plants reproduce by spores.

Conifer and cycad plants (18) appeared 300 million years ago. These plants have seeds. Flowering plants (22) evolved about 140 million years ago, in the middle of the Dinosaur Age.

Amphibians (16) moved onto the land 360 million years ago. Amphibians (which include frogs) lay small fragile eggs in the water, like fish eggs. They begin their active lives like fish as tadpoles, developing lungs and feet as they mature.

The first reptile appeared 335 million years ago (17). Reptiles have large amniotic eggs, like bird eggs, that can hatch on land.

Dinosaurs (19) and mammals (20) appeared about the same time. For 140 million years, from 205 to 65 million years ago, the dinosaurs dominated the land during the Age of Dinosaurs (21). In this period, few mammals were larger than a mouse. Birds (23) appeared in the middle of the dinosaur period, and apparently evolved from dinosaurs. Birds competed with pterodactyls, which were flying reptiles closely related to dinosaurs. Pterodactyls existed throughout the dinosaur period, and died out with the dinosaurs.

Sixty-five million years ago, a meteorite about 10 kilometers in diameter hit the earth near the Yucatan peninsula in Mexico, and caused a catastrophe that killed off the dinosaurs. Mammals and birds survived, probably because many were small and could eat a variety of foods. Without competition from the dinosaurs and pterodactyls, mammals and birds evolved rapidly, and our modern Age of Mammals (24) began.

Let us now examine in detail the stages involved in the development of life on earth. We begin with the explosion of animal life that occurred at the start of the *Cambrian* period, 545 million years ago.

The First Animals

Abundant fossils of multi-celled animals suddenly appeared in our oceans at the start of the *Cambrian* period, 545 million years ago. Within 5-10 million years an evolutionary explosion occurred, in which animals from essentially all animal types (phyla) have been identified.

The extremely fast evolutionary pace at the start of the Cambrian has long amazed scientists. Part of the explanation lies in the 60-million year *Ediacara* period, which preceded the Cambrian. Ediacara fossils are scarce because the animals had soft bodies, which rarely produced fossils. The Ediacara fossils are mud casts of the soft bodies. Some Ediacara fossils have been identified as jellyfish, soft corals, earthworms, and mollusks. However many are strange; some look like small fluid-filled air mattresses, unlike any animal or plant today. [46, 47, 55]

Severe Climates of the Young Earth

Another clue to the mystery of the Cambrian evolutionary explosion, 545 million years ago, is that two to four severe global ice ages occurred between 750 and 580 million years ago, which froze the whole earth.

To place this in perspective, let us examine the climate of the young earth. When the earth was formed 4.6 billion years ago, the sun was only 70 percent as bright as today, and 2.8 billion years ago the sun was only 80 percent as bright. With such a weak sun, one would expect ice ages. However, except for the ice ages between 750 and 580 billion years ago, and another ice age 2.3 billion years ago, the earth was warm in the pre-Cambrian period.

For 500 million years after the earth's crust hardened, 4.3 billion years ago, the earth was heavily bombarded by meteorites. The energy released by these impacts kept the earth very warm. By the time the meteorite bombardment ended, 3.8 billion years ago, archaea had evolved. The archaea fed on gasses emitted by oceanic volcanic vents, and released methane into the atmosphere as a byproduct. Since methane is a strong greenhouse gas, it kept the earth warm for 1.5 billion years, until the first ice age occurred 2.3 billion years ago. [44]

Cyanobacteria first appeared 3.6 billion years ago. Like plants. these bacteria perform photosynthesis, and release oxygen into the atmosphere as a byproduct. Eventually the atmospheric oxygen was sufficient to oxidize the atmospheric methane and eliminated most of the methane. With the greenhouse effect greatly reduced, a severe ice age occurred 2.3 billion years ago. [44]

The earth recovered from the severe ice age 2.3 billion years ago, and was warm until 750 million years ago. Then the earth experienced at least two severe ice ages until 580 million years ago.

The severe global ice ages between 750 and 580 million years ago can be explained as follows. Since snow is white, it reflects most of the sunlight back into space, and tends to make the earth colder. If glaciers get large enough, the cooling process can feed on itself, and the whole earth can freeze. Geological evidence suggests that extreme ice ages occurred in which the average earth temperature dropped to -50 °C (-60 °F). The oceans froze to a depth of half a mile and stayed in this condition for 10 million years. [48]

Volcanoes are a rich source of carbon dioxide. During rain, the carbon dioxide in the air reacts with silicates and carbonates in rock to form soluble bicarbonate compounds and silicon dioxide. This process removes much of the carbon dioxide from the atmosphere. However there was no rain during these extreme ice ages, and so a large concentration of carbon dioxide from volcanoes could have accumulated in the atmosphere. It is believed that this produced a greenhouse effect that finally melted the frozen earth, and created a very hot earth. Temperatures apparently rose until the average earth temperature reached 50 °C (120 °F). After that, the temperature gradually declined and a normal climate returned. [48]

The severe ice ages between 750 and 580 million years ago would have killed most of the life on earth. However marine life could have survived in regions around undersea volcanic vents. These vents may have melted chimneys through the ice sheets, leaving isolated ponds of clear ocean water. In these isolated pockets of life, different forms of animal life could have developed, which may have created the many phyla of animal life that became evident when the Cambrian began. [48]

This theory of the Precambrian ice ages is recent. Much more study will be needed before these ice ages are well understood. Nevertheless, it is clear that extreme climates occurred between 750 and 580 million years ago. These extreme climates probably had a strong influence on the evolutionary explosion of animal life that occurred at the start of the Cambrian period, 545 million years ago.

Development of the Fish

Soon after the Cambrian evolutionary explosion, 545 million years ago, there appeared a specimen from our own phylum, the *chordates*, which includes *primitive chordates* and *vertebrates* (animals with backbones). Primitive chordates are nearly brainless filter feeders that filter microscopic organisms from water and mud. They have a notochord nerve trunk along the body that was the predecessor of the vertebrate spinal cord. The first chordate specimen was a primitive chordate that looked superficially like a worm. [46]

About 510 million years ago, the notochord nerve trunk evolved into a cartilaginous backbone, and the first vertebrate appeared. This vertebrate was a jawless fish, which was similar to the modern lamprey. Many of the early jawless fishes were armored. [55] (p. 114)

The lamprey is a parasite that feeds off larger fish by grabbing onto the side of the fish with a mouth that acts like a suction cup. The lamprey has no jaw, but has teeth that cut into the flesh of the larger fish. Early jawless fish were probably filter feeders like their primitive chordate ancestors. Nevertheless, like the lamprey, they were much more complicated physiologically than the primitive chordates, having eyes, ears, and well developed circulatory systems.

The next milestone in vertebrate development occurred about 425 million years ago, when the first fish with jaws appeared. The jaw allowed a fish to escape from a simple filter-feeder life to become a predator or browser. The vertebrate jaw is a key feature that led to the success of the vertebrates. [49, 55].

The first jawed fish (called the *Placoderm*) wore protective armor. These primitive fish were probably much heavier than water, and spent most of the time on the bottom. By 360 million years ago, all of the armored jawless and jawed fishes had become extinct.

A new innovation in fish evolution occurred about 415 million years ago, when the first member of the shark class of fish appeared. Sharks and rays have cartilaginous skeletons, which are much lighter than bone, and so the fish is only slightly heavier than the water it displaces. This allows a shark-like fish to stay at a fixed depth with only a small upward force, which is achieved by a slow forward motion.

The shark-like fish could leave the region near the bottom and travel freely at different depths. Since it could move quickly, it did not need armor to protect itself from predators.

The next major innovation in fish development occurred about 400 million years ago with the development of the air bladder in *modern bony fishes* (*Osteichthyes*). The air bladder evolved as a modification of the intestine. Since a shark-like fish does not have an air bladder, it seems likely that the primitive Placoderm fish also lacked an air bladder.

A *modern bony fish* extracts air from the water to inflate its air bladder. Although this fish is basically much heavier than the water it displaces, the air bladder provides neutral buoyancy. This allows the fish to remain motionless at a fixed depth and thereby enables it to travel freely at various depths. With this great advantage of the modern bony fish, the primitive armored fishes, with their sluggish existence near the bottom, were forced into extinction. Today all jawed fish either have air bladders (modern bony fishes) or have cartilage skeletons (sharks and rays).

There are two groups of modern bony fish: the ray-finned fish and the lobe-finned fish. Almost all of the bony fishes today are in the ray-finned group. The lobe-finned group consists only of the lungfish and the Coelacanth, which is so rare it was believed to be extinct until rediscovered in 1938. Nevertheless, the lobe-finned fish was extremely important 360 million years ago, because it evolved into the amphibian, which in turn led to reptiles, birds, mammals, and eventually to humans.

Amphibians Invade the Land

Amphibians, which now consist of frogs, toads, newts, and salamanders, evolved from fish and moved onto the land about 360 million years ago. How did this remarkable move from water to land occur? To live on land an amphibian needed lungs and feet. How did it derive these in an evolutionary process that consisted of a series of small changes?

The air bladder, which was a critical innovation in the evolution of the fish, allowed fish to make the next major step in vertebrate evolution, the development of the lung. We see this in the modern lungfish, which has gills to obtain oxygen from water and a lung to obtain oxygen from the air. The lung was formed by modifying the air bladder.

One may ask, "Why does a fish need to breathe air?" The answer was only recently discovered by paleontologists. Amphibians apparently evolved from fish that lived in swamps. [50, 51] There is very low oxygen content in swamp water, and so these fish needed to obtain oxygen by breathing air with their lungs. We do not encounter this phenomenon today, because oxygen-poor swamps are now populated by amphibians and reptiles, rather than by fish.

Since many swamps are shallow, the fish that lived in swamps often used their fins as feet to move through the swamps. Eventually the fins developed feet-like forms, with well-defined toes. It was originally thought that these *fish with feet* were true amphibians. However, recent studies have shown that the feet were too weak to have supported the body on land, and could only have been used in the water.

An even more surprising aspect of these *fish with feet* is that some had more than five toes. All land vertebrates are called tetrapods (which means they have four limbs), and each of these limbs has five functioning toes. More than five toes can sometimes occur, but they are not functioning members. Sometimes the limbs have been lost (as in snakes) or some toes have been lost. Nevertheless, the fundamental skeletal pattern of all land vertebrates has four limbs, each with five toes.

This indicates that the first amphibian must have had four limbs, each with five toes. One of the fossil *fish with feet*, which was originally classed as an amphibian, was found to have eight toes. This fossil was clearly not an amphibian, nor could it have been the ancestor of an amphibian.

The subclass of modern bony fish that includes the coelacanth and the lungfish apparently led to the first amphibian. These fish have lobe fins with a structure that is similar to the four limbs of tetrapods.

Thus we conclude that relatives of the coelacanth and the lungfish developed the ability to live in oxygen-poor swamps. They had lungs to breathe air, and developed feet to move efficiently through shallow swamps. One of these, having five toes on each foot, developed feet that were strong enough for it to crawl onto land. This animal was the first amphibian.

Why did the amphibian move onto land about 360 million years ago? The obvious attraction was food. Plants had spread over the land 70 million years earlier. These plants were followed by millipedes and other bugs that fed on the plants. The first amphibians ate these bugs. It was only later that amphibians developed the ability to eat plants.

Spread of Plants over the Land

Let us backtrack and see how plants spread over the land. Without land plants, there could be no land animals. We have seen that multi-celled marine plants, which are called algae or seaweed, appeared about 1.8 billion years ago. Terrestrial plants evolved much later, because their physiology is more complicated. They require specialized structures to obtain and hold moisture and nutrients.

About 430 million years ago, primitive terrestrial plants moved from water onto land. These were related to *liverworts* and *mosses*. Liverworts, which are more primitive than mosses, are short leafy plants that sometimes resemble livers. Mosses and liverworts lack a vascular system for conducting fluids, and so are low plants that require abundant moisture. These plants reproduce by spores. Unlike a seed, which has many cells, a spore is a single cell.

Next came the *ferns*, *club-mosses*, and *horsetails*, which have vascular systems for conducting fluids. These plants also reproduce by spores. Modern club-mosses are often called ground pines, because they can look like miniature pine trees. However many ancient club-mosses resembled palm trees, having grass-like leaves on top of a long trunk.

Horsetails are hollow reeds, like bamboo stalks, that have leaves like pine needles growing from the joints between the reed segments. Modern horsetails are typically 2 to 6 feet tall, but some ancient species were 60 feet tall and one foot in diameter.

In the *Carboniferous* period (354-295 million years ago), giant ferns, club-mosses, and horsetails, 50 to 130 feet tall, formed great tropical jungles. These produced massive coal deposits that we now use for fuel.

Cycads and *conifers*, which have seeds, appeared about 300 million years ago. Conifers bear cones and consist of pine, spruce, cedar, fir, etc. Cycads resemble palm trees, but are biologically very different. Most cycads have become extinct, and only a few species survive today. Although cycads were originally very successful, they have been largely replaced by flowering plants.

Flowering plants evolved much later, in the middle of the dinosaur period. The oldest definite fossils of flowering plants are 120 million years old, but flowering plants probably appeared early in the Cretaceous period, which started 144 million years ago. Birds evolved about the same time as flowers. What a coincidence that *flowers* and *birds*, which so enrich our lives with colorful beauty, were created about the same time!

Flowering plants have the advantage over earlier plants that they reproduce rapidly. The noted dinosaur authority, Robert Bakker, has proposed that dinosaurs greatly aided the development of flowering plants, because they devoured vegetation so rapidly. This produced large open spaces where flowering plants could germinate and grow.

Fungi, which consist of mushrooms, molds, yeast, etc., were originally classified as plants. However fungi are radically different from plants that contain chlorophyll, and so are now placed in a separate kingdom. Like animals, fungi achieve nutrition by digesting carbohydrates, but fungi perform this digestion externally by excreting enzymes. Fungi reproduce by spores.

Lichens played a key role in the spread of terrestrial plants over the land. Lichen is a symbiotic combination of fungus and algae (or cyanobacteria). The fungus forms a body that gathers and holds water and nutrients, and the algae perform photosynthesis to produce carbohydrate food. Lichens can grow on bare rock. They gradually dissolve the rock, converting it into soil on which terrestrial plants can grow. The earliest definite fossils of lichens are 400 million years old. Lichens probably existed before that, but they live in conditions where fossilization is rare, and so the lichen fossil record is meager.

As explained earlier, carbon dioxide in rain converts the carbonates of rock into soluble compounds. This process breaks down rocks and thereby helps to form soil in which plants can grow. Glaciers grind up rocks and so are also important means of making soil.

Single-celled eukaryotes (cells with nuclei) are called *protists*, and first appeared 2.7 billion years ago. Protists consist of plant-like cells (such as diatoms), animal-like cells (such as protozoa and amoebae), and fungus-like cells. Plant-like protists are also called algae.

The Reign of the Reptiles

As stated earlier, the first amphibians moved onto the land about 360 million years ago. The first *reptile* appeared 25 million years later. Reptiles have amniotic eggs, like bird eggs, which allow them to reproduce on land. Amphibians, such as frogs, have small fragile eggs (like fish eggs) that must hatch in water.

Many of the early amphibians were much larger than modern amphibians, some being 13 feet long. Amphibians dominated the land for 45 million years until they were displaced by reptiles.

Within 10 million years after the first reptile appeared, the reptiles separated into two major groups: (1) the *Synapsid reptiles*, which evolved into mammals, and (2) the *Diapsid reptiles*, which evolved into lizards, crocodiles, dinosaurs, and birds. It is often thought that a third group, the *Anapsids*, evolved into turtles, but turtles may have descended from Diapsids. We know that the Synapsid reptile was related to the mammal because, like the mammal, each side of the skull has a single hole behind the eye socket. In contrast, a Diapsid reptile skull normally has 2 holes behind the eye socket. [52]

The Synapsids

The Synapsid reptiles (the ancestors of mammals) quickly became the dominant reptiles. The Synapsids controlled the land for nearly 70 million years, from 315 million years ago until the end of the *Permian* period, 248 million years ago. At that time a mass extinction occurred. This was by far the most catastrophic extinction since the Cambrian began and it eliminated 95 percent of all animal species. [52]

The early Synapsids were Pelycosaurs. A common example was the 10-foot long carnivore, Dimetrodon. It looked like a giant lizard with a sail-like fin on top of its body, which was probably used to control body temperature. Although an ancestor of mammals, Dimetrodon is often included in the "dinosaur" collection of children's toys. [52]

In the late Permian, the primary Synapsids were Therapsids, which, unlike Pelycosaurs, may have been warm blooded. Therapsids typically had lizard-like bodies but had teeth like mammals. Usually the body was held high above the ground, but the legs sprawled sideways like a crocodile. A large herbivore, Moschops, was 17 feet long and nearly the size of an elephant. Gorgonopsid was a vicious carnivore with large canine teeth like a saber-tooth tiger. A common herbivore, Dicynodont, had a turtle-like beak instead of teeth. [52]

Some of the Synapsid reptiles survived the Permian extinction, but in the following *Triassic* period the Synapsids were eclipsed by a group of Diapsid reptiles, called Archosaurs, which includes crocodiles and dinosaurs. A small mass extinction occurred 225 million years ago, and after that the first dinosaur appeared. A much greater mass extinction occurred 205 million years ago, at the end of the *Triassic*. [53]

The causes of the mass extinctions ending the Permian and Triassic periods (248 and 205 million years ago) are being studied. Manicouagan Lake in Quebec, Canada lies in a 100-km wide meteorite crater, formed about 210 million years ago. [54] This impact from an asteroid or comet may have caused the Triassic extinction 205 million years ago.

However, other studies relate extinctions at the ends of the Permian and Triassic periods primarily to extensive volcanic activity. [56]

Volcanoes released high levels of carbon dioxide, which produced a greenhouse effect to cause severe global warming. The high temperatures could have devastated plant and animal life. High carbon dioxide levels in the ocean at the end of the Permian were augmented by radical changes in ocean currents due to continental drift. [56]

Although the causes of the mass extinctions at the ends of the Permian and Triassic periods will be strongly debated for some time, the consequences of these extinction events are clear. They radically influenced the evolutionary development of dinosaurs and mammals.

The Dinosaurs

The dinosaur movie *Jurassic Park* has popularized the name of the *Jurassic* period, which followed the Triassic and began 205 million years ago. For 140 million years, in the Jurassic and the subsequent *Cretaceous* period, gigantic dinosaurs controlled the land. Our children know many of the dinosaur names. The largest were the house-size Brachiosaurus and Brontosaurus (also called Apatosaurus). The weirdest looking were probably Stegosaurus, with triangular bony plates along its back, and the tank-like Triceratops with three horns on its head. By far the most ferocious predator was Tyrannosaurus Rex, which appeared in the Cretaceous period.

For many years it was assumed that dinosaurs were cold-blooded, and therefore sluggish like our modern reptiles. However there is abundant evidence that they were warm-blooded, which would have allowed strenuous activity. The dinosaurs in the *Jurassic Park* movie were portrayed with this concept in mind. Dinosaur paleontologist Robert Bakker was the primary proponent of the concept that dinosaurs were warm blooded and very active.

During the dinosaur era, the air was dominated by strange flying reptiles called Pterosaurs or Pterodactyls, which were closely related to dinosaurs. Some were small, but one Pterodactyl had a wingspan of 40 to 50 feet. These flying reptiles were very efficient in flight, but were awkward on the ground. [52] The Pterosaur probably evolved about the same time as the dinosaur. However its fossils are scarce, because its hollow bones were fragile.

The first *bird* (Archaeopteryx) appeared in the middle of the dinosaur era, and apparently evolved from a dinosaur. The original function of feathers was probably for insulation, and this very effective feature was later adapted to provide flight. [57] Birds had the advantage over Pterosaurs that they are very nimble on the ground. Consequently birds were probably much more efficient than Pterosaurs in small sizes.

During the dinosaur era, the seas contained many kinds of Diapsid reptiles. This included the Plesiosaur (7 to 47 feet long), which had an oval body with four paddle-like flippers, and a neck so long it looked like a sea serpent. It was probably very effective in catching fish. The Ichthyosaurs looked like giant fish, some being 50 feet long. They were the dinosaur-era equivalent of our whale. [52]

The Slow Rise of the Mammals

Many of the Synapsid reptiles that survived the mid-Triassic extinction 225 million years ago were quite similar to mammals, and from these came the first true *mammal*, which appeared 215 million years ago. The mammals were small during the dinosaur era. Most were the size of a mouse and rarely was larger than a domestic cat. Mammals remained tiny for 150 million years until the dinosaurs were eliminated.

Two key features of a mammal are accurate tooth occlusion (so that it can chew effectively) and the ability to chew and breathe at the same time. [53] Although the mammals evolved from Synapsid reptiles, the early Synapsid reptiles, and many of the Synapsids that became extinct in the Permian and early Triassic periods, did not look very much like mammals.

Sixty-five million years ago, at the end of the Cretaceous period, the dinosaurs were suddenly driven into extinction. It is generally believed that this was caused by a meteorite (an asteroid or comet) about 10 kilometers in diameter that hit the earth near the Yucatan peninsula in Mexico. This caused a disaster that killed nearly all plant and animal life on land and in the seas. The dinosaurs, pterosaurs, and marine reptiles became extinct. However enough of the small mammals and birds survived to continue their species, along with a few cold-blooded representatives of the once-dominant reptiles.

Probably the primary reasons that mammals and birds survived the mass extinction are that many were small and could eat a variety of foods, including seeds, roots, and mushrooms. The cold-blooded reptiles that survived the extinction had the advantage that a cold-blooded animal can live for long periods with little food. As was stated earlier, dinosaurs were probably warm-blooded.

Since 1978, Professor Dewey McLean of Virginia Polytechnic Institute [58, 59] has written extensively showing that enormous volcanic activity occurred at the end of the Cretaceous period. Volcanoes flooded a million square miles in India, and a pile of volcanic lava near Bombay is 1.5 miles thick. If volcanoes could have caused the massive extinction at the end of the Permian period, it seems clear that this volcanism should have strongly affected the dinosaur extinction. This suggests that two separate disasters may have contributed to the destruction of dinosaurs at the end of the Cretaceous period. Some paleontologists believe that many dinosaurs had become extinct before the meteorite impact.

With the competition from the dinosaurs eliminated 65 million years ago, the mammals took over the land and evolved rapidly. The birds flourished also, filling ecological niches in the air previously dominated by pterosaurs. Mammals and birds moved to the seas, as whales, porpoises, seals, and penguins, taking the place of extinct marine reptiles. The mass extinction 65 million years ago ended the *Age of Reptiles* and ushered in our modern *Age of Mammals*.

Many dinosaurs at the end of the Cretaceous period were probably more intelligent then the modern reptiles (turtles, lizards, snakes, and crocodiles} that survived the mass extinction. Nevertheless the mammals were apparently much more intelligent than the dinosaurs that they replaced. It seems likely that the 150 million years of dominance of dinosaurs over mammals resulted in a large increase in mammalian intelligence. Brainpower was probably a great asset in helping a tiny mammal from being eaten by a terrifying carnivorous dinosaur.

The *primate* order consists of humans, apes, monkeys, and a primitive group, called prosimians, which includes lemurs and tarsiers. Prosimians appeared 50 million years ago, monkeys appeared 35 million years ago, and apes appeared 25 million years ago.

Apes differ from monkeys in that they have no tail, they have a much larger brain, and they move in an upright manner when in the trees. Monkeys scamper on four legs along tree branches. Modern apes consist of the gibbon and orangutan, which live in Asia, and the chimpanzee and gorilla, which live in Africa.

The Ascent of Humans

Australopithecus and the Homo Genus

About 6 million years ago, an African ape, called *Australopithecus*, developed the ability to walk upright. Climate changes had thinned out jungle trees, and this ape needed to move over appreciable distances between trees. Walking upright was a natural evolution for an ape. It had the great advantage that the ape could carry clubs and rocks in its strong arms to defend itself against predators, as chimpanzees occasionally do today. The canine teeth were much smaller in Australopithecines than in chimpanzees. [60] (p. 105) This suggests that Australopithecines did not use their teeth for fighting as chimpanzees do.

Australopithecus had the same size brain as the chimpanzee, and, like the chimpanzee, ate primarily a vegetarian diet.

About 2.5 million years ago, the first member of our own genus, *Homo Habilis*, evolved from this upright ape. He had a much larger brain than Australopithecus and had the intelligence to make crude stone tools. Along with fruits and vegetables, he ate meat regularly, and the stone tools were used to butcher the animals. Eating animals was a great biological advantage, because it is much easier to derive nourishment from meat than from plants.

About 1.8 million years ago, *Homo Habilis* was replaced by *Homo Erectus*, who had an appreciably larger brain. Homo Erectus was tall (about 6 ft) and a good runner. Fossil remains of Homo Erectus are extensive, but those of Homo Habilis are limited. Consequently we have poor knowledge of the Homo Habilis skeleton.

Although the Australopithecine could walk upright, its abdomen was similar to that of the chimpanzee. It had a conical chest, a potbelly, and no waist. In contrast, the Homo Erectus skeleton was essentially the same as that of modern man, except for the head. The Homo Erectus skeleton permitted the flexible movement to allow fast running. The Australopithecines, with their chimpanzee-like abdomens, could walk but they could not run very well. [60] (p. 195)

The Australopithecines disappeared about one million years ago. This extinction may have been more the result of competition from baboons than from Homo Erectus. Since Australopithecus did not eat meat regularly, it did not compete directly with Homo Erectus, but did compete directly with the newly emerging baboons.

Hunting Capabilities of Early Man

Some anthropologists have argued (with convincing success in their profession) that Homo Erectus could not have been an efficient hunter, and was probably an opportunistic scavenger. However, this concept seems doubtful.

All predators scavenge, including lions, and so we must assume that Homo Erectus also scavenged whenever he could. Nevertheless, scavenging alone could not have yielded a steady and reliable source of meat. Hyenas have the reputation of being scavengers, yet they are also effective hunters. Vultures live exclusively by scavenging, but can search over an enormous area to find dead animals.

We must assume that Homo Erectus was a good hunter, because he had the following great advantages in hunting:

    (1) He had the intelligence to allow effective cooperative hunting;

    (2) He could carry rocks and clubs in his strong arms to mount a deadly attack;

    (3) He could throw rocks and clubs from a distance.

    (4) He was a fast runner;

Let us examine these attributes.

Cooperative hunting. Wild dogs in Africa are very efficient predators because they cooperate when attacking an animal. With his much higher intelligence, Homo Erectus could have been a very effective cooperative hunter. For example, one Homo Erectus group could have herded animals slowly into an area where another group lay hidden in ambush. With a quick dash, the hidden group could have easily made a kill.

The chimpanzee provides a good example of the effectiveness of cooperative hunting. Chimpanzees catch monkeys by attacking as a group that surrounds a monkey in the trees. Because of this cooperative action, the chimpanzees can easily catch and kill their prey.

Wielding clubs and rocks. Chimpanzees sometimes use clubs and other objects when being pressed by a predator. However, they are not very effective in this regard, because they cannot stand firmly on two legs. In the arms of Homo Erectus, clubs and rocks would have been formidable weapons.

Throwing rocks and clubs. By throwing an object from a distance, Homo Erectus could have stunned or knocked down an unwary animal. With a quick dash, the animal could have been subdued and killed. The great skill of a baseball pitcher today may well be the result of human physical capabilities that evolved over millions of years.

Fast running. It is hard to explain why Homo Erectus developed a body that allowed efficient running if it was not used for hunting. He could not run fast enough to outrun a predator.

From Homo Erectus to Anatomically Modern Humans

Homo Habilis made crude pebble tools (called *Oldowan*) by striking one lava pebble against another. Much more complicated stone tools (called *Acheulian*) were made by *Homo Erectus*. [61]

Homo Erectus had the adaptability that allowed him to leave Africa and populate Europe and Asia. Homo Erectus was found in Indonesia soon after he appeared in Africa. [60] He probably learned to use fire at least one million years ago. [*Science News*, May 1, 2004, p. 276]

Starting about 600 thousand years ago, more intelligent beings evolved from *Homo Erectus*, which are known as *Archaic Homo Sapiens*. Their brains were larger, and the types of stone tools increased. Specimens appearing in Europe about 600 thousand years ago are also called *Heidelberg* man.

A more advanced stone tool technology (called *Mousterian*) appeared about 250,000 years ago. *Neanderthal man* appeared about 200,000 years ago. He had a heavy frame and was physically much stronger than modern man. His head enclosed a brain as large as modern man, but with a different shape. He had a sloping forehead, a heavy brow ridge, a very large nose, protruding teeth, and a receding chin. He disappeared about 35 thousand years ago. [61]

The first anatomically modern human (*Homo Sapiens Sapiens*) appeared about 100 thousand years ago. The oldest fossils were found in the Middle East and in Africa. A recent discovery suggests that anatomically modern humans may have existed in North Africa 150 thousand years ago. For at least 60 thousand years, modern humans and Neanderthals coexisted. During this period, modern humans and Neanderthals made essentially the same tools, which were utilitarian devices for butchering animals, making spears, etc. [61] (p. 122)

Then, suddenly, about 40 thousand years ago, modern humans started making extensive artistic objects from bone and stone. These objects included, for example, necklaces, pendants, and bracelets. The earliest fine work of art is a beautiful ivory carving of a horse made 32 thousand years ago from a mammoth tusk. Caves in Lascaux, France contain superb drawings of animals that are 17 thousand years old, and are accompanied by a wealth of abstract symbols. These works of art demonstrate that those who made them were as intelligent as humans are today. [61] (p. 115)

What caused this sudden and drastic change in the behavior of anatomically modern humans 40 thousand years ago? This question is probably the greatest mystery in the evolution of humanity.

The Development of Sophisticated Language

Anthropologists have long considered language to be a key element for understanding the evolution of human intelligence. Apes cannot make articulate speech sounds, because their vocal tracts do not allow it.

In humans, the vocal cords are deep in the throat. The large space above the vocal cords allows the formation of articulate speech.

The principle structures of the vocal tract are the larynx (which holds the vocal cords), the pharynx (the tube above the larynx, which opens into the oral and nasal cavities), and the tongue and lips. Basic sounds are generated in the vocal cords, which are modulated in the structures above. In apes and in newborn humans, the larynx is located high in the neck, and so the possible speech sounds are very limited. In adult humans, the larynx is located low in the neck and thereby allows a wide modulation of sounds. [62]

This characteristic of modern humans that permits articulate speech comes at a heavy price. Because the vocal cords are so low, adult humans cannot breathe and swallow at the same time. Consequently humans can choke to death when they eat. Apes can breath while they swallow, and so do not have this problem. Infant humans can breathe and swallow at the same time, because the larynx is high in the neck. This is essential to infants because it allows them to breathe while they nurse. [62]

The position of the vocal cords can be determined from fossils. Studies have shown that the vocal tracts of Homo Erectus were beginning to be modified about 2 million years ago to allow articulate speech. The larynx reached its present low position at least 600 thousand years ago, as shown in the skull of Heidelberg man. [62] Since this vocal tract modification prohibits simultaneous breathing and swallowing, it must have had a strong biological advantage. Therefore it is reasonable to assume that, from the beginning of our Homo genus, **articulate speech** was an important characteristic.

Ian Tattersall, Curator of Anthropology at the *American Museum of Natural History* in New York, has presented a new theory to explain the revolutionary change in behavior of anatomically modern humans 40 thousand years ago. This is explained in a December 2001 *Scientific American* article by Tattersall, "How We Came to Be Human", and in his associated book [62]. This change was far too rapid to be explained by physical evolution, and so he concludes that it must have been a cultural change.

Tattersall believes that this new behavioral characteristic was caused by the development of **sophisticated language**. With a sophisticated language, complex thoughts could be communicated, and this resulted in a radical change of behavior. This language achievement was probably a cultural development that applied speech and mental capabilities already present in anatomically modern humans.

This theory of Tattersall becomes clearer when we examine the vast difference between *crude but articulate speech* and *sophisticated language*. There have been cases of children who have grown up isolated from other humans and so did not learn to talk until they were over 12. They were eager to learn to talk and rapidly developed an appreciable vocabulary of words. However, they could not connect these words together to form logical sentences. They had learned to speak too late in their lives to achieve a sophisticated language.

Chimpanzees and gorillas have been taught to communicate using sign language. Although they have learned a large collection of words, they cannot tie these words together into sentences.

Crude but articulate speech would have been a strong advantage to even the early members of our Homo genus, because it would have greatly assisted the task of cooperative hunting. Communication with speech would have been a strong advantage in a hunting attack. Complex language was not required. Simple commands would have been adequate to coordinate a hunting operation.

It seems likely that members of the Homo genus up to Neanderthals and early anatomically modern humans could utter *articulate* sounds, which they used to make crude commands that consisted of separate words. This crude speech would have been adequate to satisfy the essential needs of primitive life.

The radical innovation that resulted in modern humanity may have been the development of a *sophisticated language*, which included the syntax principles for connecting words together into logical sentences. With a sophisticated language, complex thoughts could be conveyed, and this could have led to an enormous cultural expansion.

Neanderthals may have lacked the intellectual capability of achieving these new language skills, and so could not share in this cultural innovation. They did not have our high forebrain, which may be essential for a sophisticated language.

Sophisticated language developed by anatomically modern humans about 40 thousand years ago would have had practical as well as artistic advantages. It would have allowed them to compete more effectively against the Neanderthals for limited food. This may explain why the Neanderthals became extinct about 35 thousand years ago.

How the Homo Genus Evolved

Ever since Homo Erectus left Africa about 1.8 million years ago, the different groups of the Homo genus apparently stayed in sufficient contact with one another for the Homo genus to remain a single species This means that the different groups could and did interbreed. There is fossil evidence of limited interbreeding between Neanderthals and anatomically modern humans, and so these two groups apparently were of the same species.

We are faced with an enormous interval of time in comparison to recorded history. The migration of Homo Erectus out of Africa may have been no faster than 25 miles within a 25 year lifetime. Over 1000 years, this slow migration could have carried early man a distance of 1000 miles. When we consider the migrations of early man, we are not implying the rapid migrations that have occurred in recent times.

How did Homo Erectus gradually evolve to produce modern humans? Milford Wolpoff at the University of Michigan is the primary proponent of the *Multi-Regional Hypothesis*. It postulates that the slow migration of early humans, and the resultant interbreeding, caused humans to evolve together and thereby remain a single species. Advantageous genes were passed slowly from one group to another. Nevertheless, regional differences remained, and these differences form the basis for the racial distinctions that we see today.

The *Multi-Regional Hypothesis* is opposed by the *Replacement Hypothesis*, which postulates the replacement of one Homo group by another. Proponents of the Replacement Hypothesis maintain that modern humans evolved at least 100 thousand years ago (and maybe much earlier), probably in Africa. They migrated and eventually replaced all of the earlier populations.

Wolpoff argues against the Replacement Hypothesis, because anatomically modern man coexisted with Neanderthal man for at least 60 thousand years. During that period the two groups made the same stone tools. The perplexing relation between modern humans and the Neanderthal "cave men" has long been an issue of debate. [61] (p. 122) We discussed this issue at the end of Chapter 1.

A possible explanation of this coexistence of two radically different types of humans is that it occurred in Europe during ice-age periods. The Neanderthals were apparently much better adapted to cold weather than anatomically modern man. (Neanderthals may have had an insulating layer of fat under the skin and more hair on their bodies.) The Neanderthals may have lived in the cold regions, while the anatomically modern humans lived in the warmer regions. If the two groups did not live in the same location at the same time, there would have been little contact between them. This assumption could explain why they remained biologically distinct.

Applying the Tattersall theory suggests that sophisticated language may have given the anatomically modern humans a great advantage over the Neanderthals about 40 thousand years ago. If the Neanderthals were incapable of achieving a sophisticated language, they could have lost in the competition for food. Modern humans may have learned the complex skills of making tight, warm clothing, and thereby could live in the cold regions populated by Neanderthals.

Outside Europe, with its ice ages, there may have been much more interbreeding, so that the populations were more homogeneous. Consequently all of the inhabitants in the other regions may have been able to learn the new sophisticated language skills.

Here is an hypothesis of how modern humanity might have developed, which may explain the relationship between modern humans and the Neanderthal "cave men". Whether or not the reader accepts this explanation, it should at least help to illustrate the confusing aspects of this enigmatic relationship.

Was there a Divine Spark in the Development of Humanity?

Our explanation of the evolution of life on earth, from early microbes to the development of modern humanity, has applied the principles of natural selection. The natural selection process capitalized on random genetic mutations, augmenting those mutations that yielded advantageous characteristics. By this means the complex life of today could have evolved from very simple life in a countless series of small changes that occurred over billions of years.

Is that all that there was? Was any divine purpose involved in this process? With DNA analysis one can demonstrate a remarkable uniformity in the biological structure of different species. It seems impossible to argue that the cells in the bodies of humans are not biologically related to the cells in the bodies of other species.

However, this does not prove that there was no divine spark in the development of humanity. The revolution of modern human behavior that occurred about 40 thousand years ago is a phenomenon that cannot be attributed to biological evolution. Modern human behavior was created far too suddenly. We have attempted to answer this with the Tattersall theory, but even this does not explain why humans suddenly began to use sophisticated language.

Since modern human behavior developed too quickly to be explained by biological evolution, one can legitimately ask the question, "Was this development the result of Divine intervention?"

The Growth of Civilization

Regardless of the causes, modern humans suddenly began to display a very high level of intelligence about 40 thousand years ago. About 12 thousand years ago, this intelligence led to a revolution in human activity with the development of agriculture. Before agriculture, humans were limited to a hunter-gatherer way of life, which greatly restricted the population that could be supported by the land.

Agriculture began about 12 thousand years ago with the planting of grains and the domestication of sheep, goats and pigs. With agriculture, the population density could dramatically increase. About 10 thousand years ago, the first towns of appreciable size appeared. One of the two earliest known towns was at the site of the Biblical town of Jericho.

After the development of agriculture, complex civilizations evolved rapidly. The pyramids of Egypt are probably the most striking evidence of this. Those marvels of engineering were constructed nearly 5 thousand years ago. It is hard to believe that the pyramids of Egypt were as old to the Classical Greeks as those Greeks are to us.

Beyond the Earth

Now let us raise our eyes from the earth to the heavens. We leave the realm of the paleontologist and enter that of the astronomer. Like the paleontologists, the astronomers have derived a tremendous amount of information from painstaking observations. This has yielded remarkable understanding of the beautiful stars that shine in the sky, including that nearby star that we call our sun.

Geological Periods (*Millions of years ago*):

Cambrian 545-490; *Ordovician* 490-443; *Silurian* 443-417; *Devonian* 417-354; *Carboniferous* 354-295; *Permian* 295-248; *Triassic* 248-205; *Jurassic* 205-144; *Cretaceous* 144-65; *Age of Mammals* 65-present. [55]

Major Periods of Vertebrates: Jawless fishes (Ordovician); early jawed fishes (Silurian); modern fishes (Devonian); amphibians (Carboniferous); Synapsids (Permian); first dinosaurs and mammals (Triassic); dinosaurs (Jurassic and Cretaceous).

Chapter 3

Historical Foundation of Astronomy

Ancient Astronomical Observations

The Babylonians and Egyptians made the earliest accurate astronomical measurements. The Egyptians needed this information for the practical purpose of developing a calendar that would predict the annual floods of the Nile, so that irrigation and planting could be coordinated with the floods. The Babylonians had similar practical needs for astronomy. However, the primary application of astronomy by Babylonians was in the mystical predictions of astrology, using astronomical data to "predict" future events on earth. The concepts of astrology, which began in Babylonia, are still widely believed today.

These observations of Egypt and Babylonia were the foundation for extensive studies of astronomy by the Greeks. Most of our knowledge of early astronomical studies was obtained from Greek literature.

The Celestial Sphere of the Stars

To understand how the ancient astronomers thought of the heavens, look up at the night sky. Fig. 3-1 shows the apparent motions of individual stars as the night progresses. The shaded area is the horizontal surface of the ground; the black dot in the center is *you* (the observer).

The small circles are stars. Star (1) is the North Star, the only star in the sky that does not move. (The North Star moves slightly, but to simplify the discussion we ignore this motion.) Stars (2) to (6) are typical stars, which move from *East* to *West* as indicated by the arrows.

Each star moves along a circular path (which appears as an ellipse in the figure). The portion of a path below the shaded area is not shown, because that portion is below the ground. Each circular path is centered about the *Rotation Axis*, which passes through the North Star. It seems as if the stars lie on a gigantic *Celestial Sphere*, which rotates one

3. *Historical Foundation of Astronomy* 37

revolution per day about the *Rotation Axis*.

Just as a globe representation of the earth is marked by circles of constant latitude, so we can imagine that the *Celestial Sphere* of the heavens has similar "latitude" circles. Each star moves along a circle of constant "latitude" on the celestial sphere. However, for the *Celestial Sphere*, the term "declination" is used instead of "latitude".

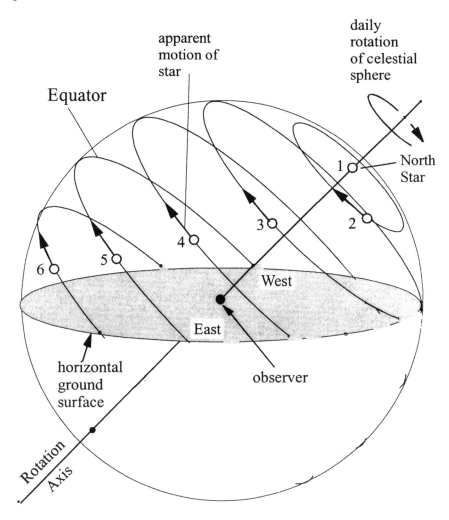

Figure 3-1: The celestial sphere representing the stars in the night sky as they move from east to west

The location of any point on the spherical surface of the earth is specified by its latitude and longitude. Similarly, the location of a star on the celestial sphere is specified by equivalent latitude and longitude coordinates, which are called "declination" and "right ascension", but right ascension is measured in terms of time, rather than degrees.

Modern astronomers use the celestial sphere concept as an artifice to describe the location of stars and their apparent motions across the sky. We know that the stars are actually fixed, and that the motion of stars across the sky is an apparent effect produced by the rotation of our earth. However, the ancient astronomers believed that the celestial sphere was a physical reality. Since the celestial sphere itself is not visible, they believed that the stars were embedded in a transparent sphere, which they called a "crystalline" celestial sphere.

One of the circles of constant declination ("latitude") on the celestial sphere is particularly important. This circle is the ***Equator***. All points on the equator are 90 degrees from the North Star. The equator is the path followed by stars that are 90 degrees from the North Star.

Only part of the celestial sphere is visible at any day of the year, because the brilliance of the sun obscures the weak light from stars during the daytime. Each year, the sun moves through a complete circle relative to the stars, and so the full celestial sphere of the stars is visible over the year. Ancient astronomers recognized that the stars are present during the daytime but cannot be seen.

The Ecliptic Path Followed by the Sun

During the year, the sun moves relative to the stars through the 12 zodiac constellations of stars. The center of the sun follows a circular path called the "ecliptic". Ancient astronomers made accurate measurements of the circular ecliptic path that is followed by the sun.

Since the sun and the stars cannot be observed at the same time, the ecliptic path of the sun was determined by indirect measurements. By timing the motion of the sun during the day, and relating this to the motion of the celestial sphere of stars during the night, ancient astronomers accurately measured the path of the sun's center relative to the stars. This was assisted by measurements between the moon and the sun, because the moon can be observed during day and night.

During the night, the stars move across the sky from east to west, and during the day the sun also moves from east to west. However, over the year, the sun moves relative to the stars in the opposite direction, from west to east.

Figure 3-2 shows a simplified celestial sphere, which includes the North Star and the *Equator*. (A partial *Equator* circle was shown in Fig. 3-1.) Figure 3-2 also shows the ecliptic path that the sun follows relative to the stars. The small circle is the sun, and the arrow is the sun's west-to-east motion over the year through the zodiac constellations of stars. The black dot in the center represents you, the observer.

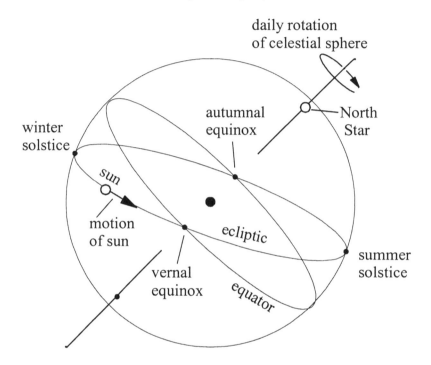

Figure 3-2: Ecliptic path of annual motion of the sun relative to the stars, and its relation to the celestial sphere equator

The plane of the ecliptic circle is tilted relative to the plane of the equator circle by 23.45 degrees. (This is the tilt angle of the earth's spin axis relative to the earth's orbit around the sun.) The ecliptic path of the sun intersects the equator of the celestial sphere at two points, which are called the *vernal equinox* and the *autumnal equinox*. The equator is the path of stars that are 90 degrees from the North Star. Therefore the vernal equinox and autumnal equinox occur when the center of the sun is 90 degrees from the North Star.

In the northern hemisphere, the sun passes through the *vernal equinox* in the spring (March 21), and through the *autumnal equinox* in the fall (September 22). The ecliptic path departs from the equator by a maximum amount at the *summer solstice* (June 21) and at the *winter solstice* (December 21). In the northern hemisphere, summer solstice is the longest day of the year, and winter solstice is the shortest day.

Development of our Calendar

An important task of early astronomers was to measure the path of the sun along the ecliptic circle in order to determine the timing of important dates. The Romans held a major holiday on the day after the winter solstice. They celebrated the fact that the days were getting longer and spring was approaching.

Julius Caesar established the first stable Roman calendar in 46 BC, which was called the Julian calendar. (That was the year 709 in Roman time, and two years before Julius Caesar was assassinated.) The Julian calendar became the basis for the calendar of Europe and eventually of the whole world.

When the Julian calendar was adopted, winter solstice occurred on December 24 and the Roman Winter Solstice holiday occurred on December 25. The last day of the year (New Year's Eve) occurred one week after winter solstice. Vernal equinox occurred on March 24.

The early Christians decided to celebrate the birth of Christ on the Roman Winter Solstice holiday (December 25). They did not know the actual date of Jesus' birth, and picked this major Roman holiday so it was not obvious that they were celebrating a Christian holiday.

The Julian calendar assumed that the solar year was exactly 365 ¼ days. Normal years had 365 days, and the fractional day was added by specifying a leap year of 366 days every four years. However, this rule does not exactly match the solar year. The Julian calendar loses one day every 128 years relative to the solar year. By 1582, the Julian calendar had lost 13 days. Winter solstice then occurred on December 11, and vernal equinox occurred on March 11.

In 1582, Pope Gregory XIII corrected the Julian calendar, replacing it with the Gregorian calendar. The Gregorian calendar added the rule that years divisible by 100 are not leap years, unless they are divisible by 400. Thus, 1600 and 2000 were leap years, but 1700, 1800, and 1900 were not leap years.

Pope Gregory added 10 days to the calendar of 1582, thereby correcting for 10 of the 13 days that the Julian calendar had lost since it was established in 46 BC. Winter solstice moved from December 11 to December 21, and vernal equinox moved from March 11 to March 21.

Pope Gregory selected this 10-day shift to match astronomical conditions at the time of the Council of Nicaea, which was held in 325 AD at the city of Nicaea in Asia Minor (modern Turkey). This council was called by Emperor Constantine to establish the accepted beliefs of Christianity, which had just been adopted as the official Roman religion. During this meeting, astronomers accurately measured the time of the vernal equinox, finding it to be March 21.

In the 371 years between 46 BC, when the Julian calendar was established, and the Council of Nicaea in 325 AD, the Julian calendar had lost 3 days. Consequently, our present Gregorian calendar is 3 days behind the Julian calendar when it was adopted. We celebrate Christmas on December 25, which is 4 days after winter solstice (rather than the day after), and we celebrate New Year's Eve 10 days (rather than one week) after winter solstice.

This story of the development of the Gregorian calendar shows how the accurate measurements of the ancient astronomers directly affected the lives of the general public.

The Path Followed by the Moon

The center of the moon moves relative to the stars through one complete cycle about every month. The moon is illuminated by the sun and displays its phases (from new moon, to first quarter, to full moon, to last quarter, back to new moon) in a period of 29 ½ days. Since the sun is also moving, the moon moves through a complete cycle relative to the stars every 27 1/3 days. Like the sun, the moon moves relative to the stars from west to east.

The moon moves through the zodiac constellations along a path that is close to the ecliptic path of the sun. However, the moon varies from the sun's ecliptic path, and follows a different path each month. Greek astronomers measured the path of the moon accurately, and from this were able to predict eclipses of the sun and the moon.

Aristotle recognized from the phases of the moon that the moon has a spherical shape. Primarily because of this finding, Aristotle concluded that the earth is probably spherical. Like Aristotle, many Greek astronomers also believed that the earth was a sphere.

The Wandering Stars

Ancient astronomers recognized that nearly all of the stars are in fixed constellations, but five stars wander relative to the fixed stars.

These five wandering stars are called *planets*, because *"planet"* is the Greek word for *"wanderer"*.

The five wandering planets were given the names of Greek gods. We know them by the names of the equivalent Roman gods: Mercury, Venus, Mars, Jupiter, and Saturn.

The planet stars wander through the zodiac constellations, following paths close to the sun's ecliptic path. However, the paths of the planets as seen from earth are very complicated. Most of the time a planet moves relative to the fixed stars from west to east, but sometimes it moves in a *"retrograde"* manner, from east to west.

The Crystalline Spheres of Pythagoras and Aristotle

In the sixth century BC, Pythagoras proposed a physical theory to explain the heavens. He postulated that the sun, moon, the five planets, and the fixed stars were attached to eight concentric crystalline celestial spheres, which rotated around the earth each day. This postulate allowed the sun, moon, and the five planets to move relative to the fixed stars.

Pythagoras postulated that the fixed stars formed the most distant sphere, and the nearest crystalline sphere carried the moon. The eight concentric spheres were arranged in accordance with the periods of motions of the bodies: The arrangement of the spheres was: [1] moon (one month), [2] Mercury (88 days), [3] Venus (225 days), [4] sun (one year), [5] Mars (2 years), [6] Jupiter (12 years), [7] Saturn (30 years), and [8] the fixed stars. The numbers in parentheses are the approximate orbital periods of the seven bodies. [24] (p. 25)

Aristotle (384-322 BC) endorsed the crystalline celestial sphere explanation of the heavens postulated by Pythagoras. Because of Aristotle's great prestige, his writings established the Pythagorean postulate as the accepted belief of the heavens through medieval times. This postulate remained supreme until it was challenged by Copernicus.

Aristotle complicated the Pythagorean concept by adding several concentric celestial spheres to provide a mechanism that controls the motions of the sun, moon, and five planets. Medieval astronomers discarded the Aristotle complexity, but usually added two spheres beyond the fixed stars to control the motions of the other spheres.

The Spherical Earth

Most Greek astronomers, including Pythagoras and Aristotle, believed that the earth was spherical. Eratosthenes (276-195 BC) estimated the circumference of the earth. In Alexandria (where he lived), he found that at summer solstice midday the sun was 7.2 degrees south of zenith, and further south, at Syene (now Aswan), Egypt, the sun was exactly at zenith. The 7.2-degree angular difference is 1/50 of a 360-degree circle. Hence Eratosthenes concluded that the circumference of the earth was 50 times the distance from Alexandria to Syene. He computed the earth circumference to be 250,000 *stadia*, where a *stadium* was the length of an Olympic stadium (estimated to be 157 to 185 meters). This gives 39,250 to 46,250 km for the earth circumference. The actual value is 40,000 km. [24] (p. 39)

The Almagest of Ptolemy

The knowledge of astronomy was greatly advanced by the works of several Greek astronomers. The greatest was Hipparchus, who lived about 150 BC. We have learned of his achievements primarily from Ptolemy, who lived three centuries later. Most of the original writings of Hipparchus have been lost.

We have discussed the ecliptic path followed by the sun, and the much more complicated paths of the moon and the planets. These complicated paths were accurately characterized by Hipparchus.

A modern mathematician could describe these paths in terms of algebraic formulas, but the Greeks did not know how to use algebra. Algebra was invented many centuries later by the Arabs. Greek mathematicians understood geometry, and so they expressed the motions of astronomical bodies in terms of geometry. [24] (pp. 41-43)

The simplest geometric description of a non-circular orbit used by Hipparchus was the eccentric circle, in which the astronomical body moved at a constant rate around a circle, the center of which was offset from the earth. This was used to represent the motion of the moon, to account for the variation of the earth-to-moon distance. The Greeks did not have the mathematical means to specify an ellipse.

Hipparchus proposed the epicycle to describe the complex motions of the planets. As shown in Fig. 3-3, in an epicycle orbit, the planet moves in a circular path about a virtual center, which in turn moves in a circular path around the earth. The epicycle could describe with reasonable accuracy the complicated planetary motions, including the intervals during which the planets move from east-to-west in the retrograde direction (backward) relative to the fixed stars.

An extensive treatise on astronomy was published about 140 AD by the Greek astronomer Claudius Ptolemaeus (c. 100-170 AD), who lived in Alexandria. He is commonly known as Ptolemy. His treatise consisted of 13 books, and is commonly known as the *Almagest*. Ptolemy used a Greek title, meaning *"Great Composition"*. Arab translators expressed this title as *Al Migisti*, which gave us the name *Almagest*. [24] (pp. 62-64)

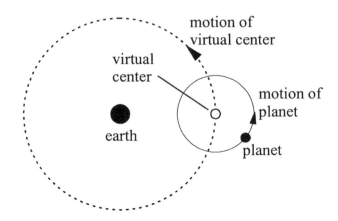

Figure 3-3: The epicycle used by Ptolemy to characterize the complicated motion of a planet

According to Ptolemy, his *Almagest* was based largely on the work of earlier astronomers, particularly Hipparchus. Ptolemy documented the quantitative observations of the earlier astronomers, and developed better geometric descriptions to characterize the data, making strong use of the epicycle.

Ptolemy assumed that the earth was spherical. He accepted the estimate of the earth circumference calculated by Posidonius (135-51 BC), which was 180,000 stadia or possibly 240,000 stadia. This was reasonably consistent with the value of 250,000 stadia obtained earlier by Eratosthenes (discussed on page 43). [24] (p. 64)   All of this information was apparently lost by the time of Columbus.

Ptolemy recognized that much of the astronomical motions could be explained by assuming that the earth is spinning. However, he rejected the concept that the earth is moving. In the words of author Berry [24] (p. 65), Ptolemy "concluded - - that it is easier to attribute motions to bodies like the stars, which seem to be of the nature of fire, than to the solid earth, and points out also the difficulty of conceiving the earth to have a rapid motion of which we are entirely unconscious".

Ptolemy's descriptions of the motions of the moon, sun, planets, and fixed stars around the earth were purely mathematical representations to express the astronomical data accurately. Ptolemy did not include the crystalline celestial sphere concept of Pythagoras and Aristotle in his Almagest. However, the medieval astronomers slavishly followed the great wisdom of Aristotle, and mixed Aristotle's crystalline sphere speculations with the accurate astronomical data presented by Ptolemy.

Popular Acceptance of the Spherical Earth

Although many Greek astronomers, including Ptolemy, believed that the earth was spherical, the general public did not. Except for mountains and valleys, the earth superficially seems to be flat. Consequently, nearly everyone in the ancient world believed that the earth was flat.

As sailors became more proficient in navigating their ships, they came to realize that the earth is spherical. They used the North Star to give the direction of north. As the mariners traveled north, they discovered that the North Star moved up higher in the sky. This effect became large as the ships began to travel from the Mediterranean to northern Europe. The angle of the North Star above the horizon told the navigators the latitude of the ship, which describes location in the north-south direction.

The change of the angle of the North Star above the horizon as one travels in the north-south direction proved that our earth was spherical. Consequently, by the time of Columbus, all competent navigators knew that the earth was a sphere; it was round like a ball. Since the earth is round, one could theoretically reach the Spice Islands in the East Indies (Indonesia) by traveling west. However, the distance that must be traveled in the westerly direction to reach the East Indies was believed to be far too great to make such a voyage practical.

Columbus thought otherwise. It was difficult to measure distances in the east-west direction, and Columbus thought that the East Indies were much further in the east than most people believed. He also thought that the earth was smaller than was generally believed. He probably heard stories of Viking explorations showing that there was land to the west that could be reached by his ships. He apparently thought that this land was Asia, but it was the unknown continent, America.

The voyage of Magellan, which was completed in 1522, circumnavigated the earth. This definitely proved to everyone that the earth was round.

The Copernicus Revolution in Astronomy

Our concept of the heavens was revolutionized in 1543 by the astronomical theory of Nicholas Copernicus (1473-1543). He was a clergyman (possibly a priest), secretary, and physician at the Frauenburg Cathedral in Polish East Prussia. His Polish name was Mikolai Kopernik.

Copernicus was well educated, and was particularly interested in mathematics and astronomy. He had the rare ability to reason for himself, and he recognized that the accepted model of astronomy did not make sense. He knew that the earth was a sphere. If the spherical earth is spinning like a top at one revolution per day, the heavens would appear to move across the sky each day. This assumption would enormously simplify the explanation of astronomy.

Greek astronomers had argued that if the earth were spinning it would have to move at such a great speed it would fly apart. However Copernicus reasoned that if the fixed stars are spinning around the earth at their enormous distances, the forces on them would be very much greater than the forces on a spinning earth.

If the earth is spinning at one revolution per day, one can assume that the fixed stars are very distant objects that do not move. Since the sun moves through the zodiac constellations of stars at one cycle per year, one may assume that either the sun is revolving around the earth at one cycle per year, or the earth is revolving around the sun at that rate. The moon receives its illumination from the sun, and moves relative to the fixed stars at one cycle per month. This suggested that the moon revolves around the earth at one cycle per month.

The data in Ptolemy's *Almagest* gave the motions of the planets relative to the fixed stars. When Copernicus assumed that the earth and the planets revolve around the sun, he found that most of the very complicated motions of the planets given in the *Almagest* could be explained by assuming that the planets and the earth are moving in circular orbits around the sun. The Copernicus assumption provided a remarkably simple picture of the heavens.

Mercury is never more than 29 degrees from the sun, and Venus is never more than 47 degrees. Copernicus concluded from this that Mercury and Venus must be closer to the sun than the earth, with Mercury being the closest to the sun. Copernicus assumed that the planets move continuously in their orbits, and that their apparent retrograde motions are caused by the motion of the earth in its orbit around the sun. With this assumption, he was able to determine the relative diameters of the planetary orbits.

To describe the motions of the planets with high accuracy, Copernicus applied Ptolemy's epicycle concept. It remained for Kepler to characterize all of the planetary orbits accurately with ellipses.

The astronomical theory of Copernicus conflicted with traditional beliefs that had been held for thousands of years. Consequently, it was strongly rejected by most people of his time. Copernicus released a preliminary version of the theory, which received harsh criticism from many sources, including Protestant leaders Luther and Calvin. However, Pope Leo X expressed open-minded interest. [66]

Copernicus completed the full manuscript of his theory in 1530, but withheld publication because he was afraid of the opposition it would receive. He agreed to let friends publish it when he knew he was close to death. He examined a copy of the printed book on his deathbed. [66]

The Copernicus document did not evoke the controversy that one might expect, because it was packed with astronomical calculations, and so could not be understood by the general reader. Besides, a friend of Copernicus added a preface saying that the book contained "merely abstract hypotheses, convenient for purpose of calculation".

We often look with disdain at Ptolemy's astronomy, but it was the quantitative astronomical data in Ptolemy's Almagest that gave Copernicus the foundation for his treatise.

It took a century for the Copernicus theory to be widely accepted. The theory was greatly advanced by the research of Kepler and Galileo.

The Planetary Laws of Kepler

Tycho Brahe (1546-1601) was a Danish astronomer who made accurate measurements of the motions of the planets. With very large instruments having no optical magnification, assisted by young observers with good eyesight, Brahe made astronomical observations over many years to very high accuracy. When Tycho Brahe died suddenly in 1601, Johannes Kepler (1571-1630) became director of Brahe's observatory near Prague.

Kepler did not have the funding to perform new observations, but he possessed the accurate observational data obtained by Brahe. Kepler applied these accurate data to the Copernicus model, and discovered that the planets are moving in ellipses as they orbit the sun, rather than in circles. From his calculations, Kepler derived three accurate relations for the orbits of the planets, which are known as Kepler's Laws. These are

(1) A planet orbits the sun in an elliptical orbit, with the sun at one focus of the ellipse.

(2) A planet moves more rapidly when nearer the sun than when further away, such that a radius drawn from the planet to the sun sweeps over an equal area for an equal time interval.

(3) The expression r^3/T^2 is the same for all planetary orbits, where r is the mean distance of the orbit from the sun, and T is the period of the orbit.

Kepler published the first two laws in 1609. He published all three laws in 1621 in his *Epitome of Copernican Astronomy*, which was Kepler's major publication. It was the first astronomical textbook based on the Copernican system, and was the primary source of information on the subject for 30 years.

Kepler died in poverty in 1630. Because of war, Kepler was unable to collect the arrears of his Imperial salary. Another personal blow to Kepler was that his mother was imprisoned 13 months for witchcraft, and died in 1622 soon after her release. [67]

Galileo's Telescope Observations of the Heavens

In 1609, Galileo Galilei (1564-1642), an Italian professor at the University of Padua, near Venice, learned of the telescope that had been invented in the Netherlands, and began making his own telescopes. In 1610 Galileo pointed his telescope toward the heavens and made observations that revolutionized astronomy.

For hundreds of years, optical lenses had been made to correct visual defects. About 1608 a Dutch spectacle maker (Hans Lippersheim or Jacob Metius) built the first telescope, which had a magnification of 3 to 4. It consisted of a convex objective lens and a concave eyepiece. (A convex lens is curved outward, whereas a concave lens is curved inward, like a *cave*.)

Galileo heard of this and began grinding lenses to make better telescopes. In 1609 he supplied the governor *(doge)* of Venice with an 8-power telescope. This was so valuable for naval use that Galileo's salary was doubled, and he received lifelong tenure as a professor. In 1610 Galileo received the appointment as the "First Philosopher and Mathematician" to the Grand Duke of Tuscany, and he moved from Venice to Florence, Italy.

Galileo built a 20-power telescope, which he directed at the heavens. In 1610 Galileo revolutionized astronomy with his telescope observations. He studied our moon and found that it had mountains and valleys like the earth. He looked at Jupiter and made the fantastic discovery that Jupiter has moons of its own. He saw that Venus has phases like our moon, which showed that Venus must revolve around the sun. He found that the pale band of the Milky Way consists of countless stars. Galileo published his findings, and claimed that his astronomical observations proved that the Copernicus theory must be correct.

Galileo received strong criticism for his support of the Copernicus theory from a number of influential university professors and clergymen. Galileo contributed to this conflict by writing material that was taken as a personal affront by politically powerful intellectuals. They finally convinced the Roman Catholic Church to bring Galileo to trial. In 1633 Galileo was convicted of heresy and imprisoned.

However Galileo was soon placed under house arrest, and lived comfortably in his villa near Florence. He was able to receive visitors and teach students. He wrote defiant books that were smuggled out to foreign publishers. He died in 1642. [67, 69]

The Galileo incident is often considered to be part of a continual conflict between science and Christianity, particularly the Roman Catholic Church, but this concept is simplistic. Galileo's book *Dialogue*, published in 1632, infuriated powerful intellectual leaders. These intellectuals used their political power to have Galileo's voice suppressed by the government. In Florence, Italy at the time of Galileo, the Catholic Church was the government.

In medieval times the Roman Catholic Church was the center of learning in Europe. Astronomy was developed by the Greeks, not by the Hebrews, and is not directly related to Christianity. However, the professors in the Christian universities found that the astronomical concepts of Copernicus and Galileo directly challenged their authority. Consequently, they used their great influence to attack Galileo.

Between 1450 and 1700, thousands of people were executed throughout Europe for witchcraft, in both Protestant and Catholic areas. When compared with this, Galileo's punishment was not as harsh as it may seem today.

Galileo's Measurements of Falling Bodies

A problem faced by the Copernicus theory is that points on the earth must be traveling at very high velocities in order for the earth to spin at one revolution per day. Skeptics of the theory asked, "How can we be moving at such a great speed without feeling the motion?"

The circumference of the earth is 24,900 miles. Hence a point on the equator moves about 24,000 miles in 24 hours, or about 1000 miles per hour. In Italy, where Galileo lived, the rotational speed of the earth is about 700 miles per hour.

Galileo realized that we are not aware of the great rotational speed of the earth, because everything around us is moving at the same velocity. To demonstrate this point, he performed several experiments with falling objects.

While riding a horse, Galileo dropped a ball. From his location in the horse, the ball seemed to fall straight downward. However, to an observer on the ground, the ball followed a curved trajectory. The same effect was observed when dropping a ball within the cabin of a ship. The ball dropped straight downward within the cabin, regardless of whether the ship was stationary or moving at high speed.

Galileo measured the motions of falling objects and discovered that, when air resistance is negligible, all objects fall at the same rate regardless of size or weight.

Galileo found that the rate of change of velocity for a falling object is constant. The rate of change of velocity is called acceleration. The velocity of a falling object increases at a rate of about 10 meter/sec (meters per second) for every second. In one second the velocity is 10 meter/sec; in 2 seconds it is 20 meter/sec; in 3 seconds it is 30 meter/sec; etc.

We call the acceleration of a falling object the "acceleration of gravity", which is denoted "g". (The exact value of the acceleration of gravity is 9.8 meter/sec for every second.)

Galileo measured the motions of objects rolling down an inclined plane. Since rolling motion is slower than that of a falling object, Galileo was able to prove with greater accuracy that the rolling bodies moved down the plane with constant acceleration.

The Development of the Telescope

The telescope that Galileo used to perform his astronomical observations was crude. One of its problems is that it had a narrow field of view. Kepler invented a much better telescope for astronomy, which used an eyepiece with a convex lens, rather than a concave lens. The Kepler telescope gives a much wider field of view, but the image is inverted. The inverted image is no problem in astronomy, but makes the telescope unsuitable for terrestrial use.

Modern terrestrial telescopes generally use the Kepler design, with an additional lens to re-invert the image. High quality binoculars use the Kepler design with prisms to re-invert the image. Inexpensive "field-glass" binoculars use the Galileo design with a concave eyepiece, and so have a narrow field of view.

The most serious limitation of the telescopes used by Galileo and Kepler is that they had chromatic aberration, which produces color fringes around the image. This defect is similar to the effect of passing light through a prism, which separates the wavelengths of light into a rainbow pattern.

Today chromatic aberration can be eliminated by using achromatic lenses, which have two kinds of glass with compensating spectral refraction characteristics. Achromatic lenses were not available to Galileo and Kepler. The achromatic lens was invented by optician John Dollard in 1757.

Without an achromatic lens, the telescope that Galileo used to observe the planets was very poor. Although Galileo could see that Venus has phases like the moon (which proved that Venus revolves around the sun) the shape of the Venus image was very difficult to discern with the crude Galileo telescope.

A half century after Galileo performed his telescope observations, Isaac Newton (1642-1727) made a great advance in telescope design by building the first reflecting telescope in 1668. This had a concave spherical mirror to concentrate the light rays, rather than a lens. All wavelengths are reflected from a mirror at the same angle, and so the major optical element in a reflecting telescope has no chromatic aberration.

The spherical mirror in Newton's telescope was made from speculum metal, which is an alloy of tin and copper. Speculum metal was used to make the mirrors of reflecting telescopes until the late 1800's, when it was replaced by glass with a silver coated surface.

Newton's Theory of Gravity

A few years after inventing the reflecting telescope, Newton discovered the principles of calculus. He applied his new mathematics to develop his Theory of Gravity.

Newton learned from Galileo that objects fall with a constant acceleration, which was a key principle of Newton's theory. Newton applied his theory to explain the orbits of planets around the sun, and found that his theory accurately satisfied Kepler's three laws. Thus, the research by Galileo and Kepler were fundamental elements in the development of Newton's theory.

Newton demonstrated that the motions of the planets around the sun, and the motion of the moon around the earth, can be accurately explained by the following two laws of mechanics:

> *(1) Law of gravitational attraction: Two bodies are attracted together with a force that is proportional to the product of their masses, and is inversely proportional to the square of the distance between their gravitational centers.*
>
> *(2) Law of motion: The force applied to a body is equal to the mass of the body times its acceleration, where acceleration is the rate-of-change of velocity.*

Newton also gave the following corollaries of his two basic laws:

> *(3) A body at rest, or moving at constant velocity, stays in that condition unless a force is applied to it.*
>
> *(4) For every action there is an equal and opposite reaction.*

By applying calculus to these laws, Newton was able to calculate accurately the motions of bodies in our solar system. Newton's theory was published in 1687 by the Royal Society of England as *Philosophiae Naturalis Principia Mathematica*. Latin was used because scientists in all countries could read Latin. The famous scientist, Edmund Halley (1657-1742), personally paid for the printing. An English translation of Newton's *Principia* is given in Ref. [29].

Newton's invention of calculus was the key that allowed Newton to develop his physical laws. Nevertheless, he did not use calculus in his *Principia*, because other scientists did not understand it. Instead, Newton applied graphical constructions that achieved the effect of calculus.

It is commonly believed that Newton discovered the principle of gravitational attraction, but this is not so. Forty-two years before Newton published his *Principia*, Ismaelis Bouillard had postulated that mutual attraction of the planets varies inversely as the square of the distance between them. [68] Besides, even before that, Galileo must have understood that gravitational attraction holds the planets in their orbits around the sun, and holds the moon in its orbit around the earth.

What Newton actually developed was a precise set of mathematical laws, with the proof that these laws accurately describe the orbits of bodies in our solar system. Without his invention of calculus, Newton could not have developed his theory.

How Cavendish Weighed the Earth

The constant of proportionality in Newton's law of gravitational attraction is the *gravitational constant* G. The formula for this law is:

Gravitational force $= GM_1M_2/(d_{12})^2$

In this equation, M_1 and M_2 are the masses of the two bodies, and d_{12} is the distance between the centers of the bodies. The expression $(d_{12})^2$ is the square of d_{12}, and means d_{12} times d_{12}.

Newton did not know the actual value for the gravitational constant G. This constant was first measured by Henry Cavendish (1731-1810). Cavendish was a brilliant and wealthy scientist, but was reclusive and eccentric. He is best known for his chemistry research, which included the discovery that water is a compound of hydrogen and oxygen.

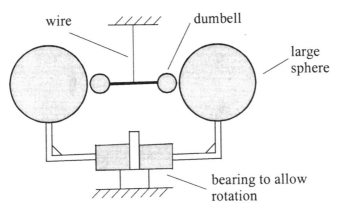

Figure 3-4: The Cavendish experiment to measure Newton's gravitational constant G

In 1798, Henry Cavendish measured the gravitational constant G by sensing the gravitational attraction between lead spheres. The experiment by Cavendish is shown in Fig. 3-4. A horizontal dumbbell, with lead balls at each end, was suspended from a thin vertical wire. A pair of much larger lead spheres was mounted on a rotatable structure, which moved the spheres just outside the lead balls. As the large spheres moved past the balls of the dumbbell, the gravitational attraction between the dumbbell balls and the large spheres made the dumbbell rotate, and caused the wire to twist slightly. The amount of twist was sensed optically to give a measure of the gravitational attraction.

The weight of a dumbbell ball was about 10 million times greater than the gravitational force between the ball and the large sphere. Nevertheless the experiment was sufficiently sensitive to detect this tiny gravitational force. From this experiment, Cavendish measured the gravitational constant G within an error of 1.4 percent.

The acceleration of gravity g is equal to (GM_e/r_e^2), where M_e and r_e are the mass and radius of the earth. Hence M_e is equal to (gr_e^2/G). Newton knew the values of g and r_e, but did not know the value for G. The measurement of the gravitational constant G allowed Cavendish to calculate the mass of the earth M_e. Consequently it was said that, *"Henry Cavendish weighed the earth in his experiment"*. He found that the average density of the earth is 5.5 times that of water.

Orbit of the Earth around the Sun

Let us see how Newton's theory explains the motion of the earth around the sun. The earth orbit is close to circular. The earth orbits the sun at a nearly constant velocity of 30 km/sec (kilometers per second).

You can illustrate the motion of the earth in its orbit by swinging a ball on the end of a string. You are the sun and the ball is the earth. The tension in the string represents the gravitational attraction between the sun and the earth. The force that the string applies to the ball is the gravitational force that holds the ball (earth) in a circular trajectory. If you let go of the string, the ball flies away in a straight line (except for the downward drop due to gravity). Thus the gravitational pull from the sun keeps the earth in its orbit, and prevents it from flying off into space.

The gravitational force that the sun applies to the earth causes the earth to accelerate. However this acceleration does not change the value of the earth velocity (which we call the *speed*), it merely changes the direction of the earth velocity.

This concept is illustrated in Fig. 3-5. The earth is shown in two locations (1 and 2) in its circular orbit around the sun. At location (1) the earth velocity is represented by the arrow V_1, which is called a "vector". At location (2) the earth velocity is represented by the vector arrow V_2. The direction of an arrow shows the direction of the velocity of the earth at that point.

In diagram (b), arrows V_1 and V_2 are moved in parallel directions, so that they start at the same point. Vector ΔV ("delta V") is the difference between these two arrows, and is constructed by drawing an arrow from the tip of V_1 to the tip of V_2.

The Greek letter Δ (delta) is commonly used to denote a difference. In this case ΔV is used to denote a difference in velocity V.

The length of the velocity difference arrow ΔV is the amount by which the earth velocity changes as the earth moves from point (1) to (2). Since the earth velocity is continually changing, the earth is accelerating as it rotates around the sun. However, this acceleration does not change the speed of the earth in its orbit, which is 30 km/sec. The acceleration only changes the direction of the earth velocity.

The earth is continually accelerating, which means that it is continually falling toward the sun. However, the earth does not get any closer to the sun, because it is moving so fast. The earth velocity moves the earth away from the sun as fast as it falls toward the sun.

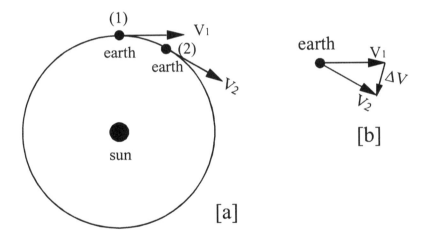

Figure 3-5: Change in the velocity of the earth as it rotates around the sun

Newton's fourth principle states that *for every action there is an equal and opposite reaction*. Hence the earth applies a gravitational force to the sun that is equal and opposite to the gravitational force that the sun applies to the earth. The gravitational force applied by the earth to the sun causes the center of the sun to wobble slightly. In recent years astronomers have measured this effect on stars to detect the presence of planets revolving around the stars. With this method, they have detected several large planets the size of Jupiter. [38] They have recently detected a planet the size of Neptune.

Why Are Astronauts Weightless?

Why do astronauts in space experience weightlessness? Although this phenomenon is well known, many people do not understand why it occurs.

The gravitational force on an object is called its weight. The weight of an object is equal to its mass multiplied by the acceleration of gravity. The acceleration of gravity is proportional to $1/r^2$, where r is the distance to the center of the earth. For a space vehicle in low earth orbit, the distance r is not much greater than at the surface of the earth, and so the acceleration of gravity is not much less. Hence, the gravitational force on the astronaut in space is nearly the same as when he is on the ground. Nevertheless, the astronaut is weightless in space. Why?

The reason for the weightless condition is that the space vehicle and the astronaut are in a free-fall state. They are continually accelerating toward the center of the earth with the acceleration of gravity g that corresponds to that altitude. However, the vehicle is traveling so fast that the altitude of the space vehicle stays constant. The acceleration merely changes the direction of the velocity.

The *International Space Station* has an altitude of 418 km (260 miles). At this altitude, the acceleration of gravity is 88 percent of the value at the surface of the earth. The space station stays at a constant altitude if its velocity V is 7.67 km/sec (4.77 mile/sec).

The space vehicle and the astronaut are continually falling toward the center of the earth, even though their altitude does not change. In this free-fall condition, the astronaut is weightless.

A sky-diving parachutist experiences weightlessness for a few seconds after jumping out of an airplane, until the velocity is sufficient for air resistance to offset the force of gravity. The wind force from the air is proportional to the square of velocity. It matches the sky diver's weight at a speed of about 120 mph when the sky diver's body is oriented horizontally, or 180 mph when the body is oriented vertically. During the first few seconds the wind force is very small, and so the falling sky diver is essentially weightless.

A simple way to experience temporary weightlessness is to jump from a high platform into water. If the height is 16 feet (5 meters), it takes one second to reach the water. During that one-second period, the individual is weightless.

The author experienced a drop of about 16 feet into water. For one second I had no feeling of gravitational force on my body, which was a very strange sensation.

The Independent Invention of Calculus by Leibniz

By demonstrating that his laws accurately explain the motions of the planets and our moon, Newton established the validity of his laws. This proved that his laws also apply to the motions of objects here on earth, and so his laws became powerful engineering tools for practical applications. In the 1600's, advances in mechanical equipment had stimulated the search for basic scientific knowledge. Engineers needed to know how to build better mechanisms, and Newton's laws had immediate practical use in engineering applications.

The famous and brilliant German mathematician, philosopher, and statesman Gottfried Wilhelm Leibniz (1646-1716) invented calculus independently of Newton. He discovered calculus in 1675, nine years after Newton, but he published it in 1684 before Newton did. With the help of the Swiss mathematicians Jacques Bernoulli (1654-1705) and his brother Jean Bernoulli (1667-1748), the Leibniz calculus concepts were refined into a convenient scientific tool that became widely used.

It was the combination of this calculus tool developed by Leibniz and the Bernoulli brothers, along with Newton's laws, that gave the world a practical engineering approach to mechanics.

Another scientific achievement of Leibniz was his invention in 1672 of a calculating machine for multiplying, dividing and taking square roots.

Newton's Revolutionary Research on Optics

Newton revolutionized the science of optics as well as the science of mechanics. In his book *Opticks* [30], Newton proved that light has a continuous spectrum. As explained in *Believe* [1] (Chapter 3) and in *Story* [2] (Chapter 6), this principle (so obvious to us today) was a revolutionary concept in Newton's day. Cohen [70] explains that Newton's optical research was rejected until he became famous for his theory of gravity. Newton's book *Opticks,* published in 1704, set the foundation for tremendous advances in optical science. Nearly all of the material had been written 35 years earlier.

Chapter 4

The Solar System and the Stars

Our Solar System

When we look at the night sky with our naked eye, we see five wandering stars, which we call planets. The brightest is Venus, named after the Roman goddess of love, because this planet is so brilliant and beautiful. The red planet is named after Mars, the god of war. The planet that moves the fastest is named after Mercury, the messenger of the gods. Jupiter, which is bright and moves slowly and ponderously, is named after the emperor of the gods. The planet that moves the slowest is named after Saturn, the god of time (equivalent to our "father time").

To the naked eye, the wandering "planet" stars look like points of light, just like any other star. However, when Galileo observed these planets with his telescope, he saw them as disks. They are unlike true stars, which look like points of light, even to the largest telescopes. We see the disk of a planet because it is very much closer than a star. The nearest star that can be seen by the naked eye is Alpha Centauri. This star is 10,000 times further away than Neptune, the most distant major planet, and 30,000 times further away than Saturn, the most distant planet visible to the naked eye.

When astronomers began to examine the sky with powerful telescopes, they discovered three other planets. Uranus was discovered in 1781, and Neptune was discovered in 1846, based on predictions derived from perturbations in the Uranus orbit. Finally, the most distant planet, Pluto, was discovered in 1930.

Characteristics of the Planets

Table 4-1 lists the major characteristics of the nine planets, including the earth. The characteristics of the sun are shown for comparison. Additional characteristics are shown in Table 4-2. Data in these tables were obtained from Ref. [27], Table 1, p. 479.

Column (1) shows the distances of the planets from the sun, relative to the distance of the earth. (The earth is 150 million kilometers from the sun.) This gives the average radius of each elliptical planetary orbit. The first planet, Mercury, has an orbital radius that is 0.39 times the earth orbital radius. The outermost planet, Pluto, has an orbital radius that is 39 times the earth orbital radius. Thus, the outermost planet is 100 times further from the sun than the innermost planet.

Table 4-1: Major characteristics of the planets and the sun.

Planet	Relative Distance from Sun	Orbit Period	Relative Diameter	Relative Mass	Density	Distance Ratio
Mercury	0.39	88 d	0.38	0.06	5.4	
Venus	0.73	225 d	0.95	0.82	5.2	1.81
Earth	1.00	365.26 d	1.00	1.00	5.5	1.37
Mars	1.53	1.88 y	0.53	0.11	3.9	1.53
Jupiter	5.2	11.9 y	11.2	318	1.33	3.47
Saturn	9.5	29.5 y	9.5	95	0.69	1.83
Uranus	19	84 y	4.0	14.5	1.3	2.01
Neptune	30	165 y	4.0	17.2	1.5	1.56
Pluto	39	249 y	0.18	0.003	2.2	-----
Sun			109	333,000	1.4	
	(1)	(2)	(3)	(4)	(5)	(6)

Column (2) gives the period of the planet orbit. Mercury orbits the sun in 88 earth days; the earth orbits the sun in 365.26 days, Mars orbits the sun in 1.88 earth years, Neptune orbits the sun in 165 earth years, and Pluto orbits the sun in 249 earth years.

Column (3) gives the diameters of the planets relative to the diameter of the earth. (The earth's diameter is 12,756 km.) Venus and Earth have nearly the same diameter. The diameter of Mars is 53 percent of the Earth's diameter. Jupiter, Saturn, Uranus, and Neptune are gaseous giants that are much larger than the earth, having diameters that are 4 to 11 times the diameter of the earth. Mercury has 38 percent of the earth's diameter. However the smallest planet is Pluto, which has only 18 percent of the Earth's diameter. The diameter of the sun is 109 times the diameter of the earth.

Column (4) gives the mass of each planet relative to the earth. Venus has nearly the same mass as the earth. Since Venus is only slightly closer to the sun than the earth, it was long considered to be our "sister" planet.

However, we now know that Venus has an enormous greenhouse effect, with global warming so great that lead would melt on its surface. We will consider what happened to Venus later.

Mars has only 11 percent of the mass of the earth. Consequently its gravity is much less. With its weak gravity, it is not able to hold a heavy atmosphere, nor an ocean of water. Consequently, Mars is an inhospitable desert, but is probably our only hope for life of any kind in our solar system.

Jupiter is by far the largest planet of the solar system. It has 318 times the mass of the earth. Saturn has nearly 100 times the mass of the earth. Neptune and Uranus have about 15 times the mass of the earth. Pluto has only 0.3 percent of the earth mass. Being so small, it is entirely unlike the other outer planets. Many astronomers believe that Pluto came from the Kuiper belt of comets that orbits outside the planets. We will discuss the comet belts later.

The sun has 333,000 times the mass of the earth. It has 1000 times the mass of Jupiter, the heaviest planet. The sun has 99.87 percent of the total mass of the solar system.

Column (5) gives the densities of the planets and the sun. The four inner planets are rocky and have high densities. Mars has the lowest density (3.9) of the inner planets. Rock has a density of about 3. The earth density is much higher than 3, because the earth has an iron core.

The major outer planets are largely gaseous, and so have low densities. Saturn is so light it could float on water. The density of water is 1.00 and Saturn's density is 0.69. The density of Pluto is only 2.2, even though it is solid. (Densities for Neptune and Pluto were obtained from Ref. [72], page 22.)

The average density of the sun is 1.4. The sun's density varies from 200 at its center to about 0.001 at its surface, which is about the density of the earth's atmosphere.

Column (6) shows the ratio of the orbital radius of each planet, to the orbital radius of the next inner planet. The ratios are calculated from the data in column (1). This shows that the Venus orbital radius is 1.81 times the Mercury orbital radius; the earth orbital radius is 1.37 times the Venus orbital radius; the Mars orbital radius is 1.53 times the Earth orbital radius; and the Jupiter orbital radius is 3.47 times the Mars orbital radius.

The ratios of column (6) are all nearly the same except for the ratio for Jupiter, 3.47, which is much greater than all the rest. This suggests that there should ideally be another planet between Mars and Jupiter. If the radius of this missing planet were 1.86 times the Mars orbital radius, the orbital radius of Jupiter would be 1.86 times the orbital radius of the missing planet. The missing planet would have a radius that is 2.8 times the radius of the earth orbit. This is about the average orbital radius of the asteroids.

The Asteroids

Between the orbits of Mars and Jupiter, there are about one million small bodies orbiting the sun, which are called ***asteroids***. They can be very harmful to our earth, because some of them have been perturbed from the normal asteroid orbits (called the asteroid belt) into orbits that intersect the earth orbit. If an asteroid should collide with the earth, even a very small asteroid, it would cause catastrophic damage.

Most asteroids are in orbits from about 2 to 3 times the earth radius. The largest asteroids are in this region and are as follows, along with their diameters: ***Ceres*** (933 km), ***Pellas*** (523 km), ***Vesta*** (501 km), ***Hygiea*** (429 km), ***Eunomia*** (272 km), ***Juno*** (267 km), ***Euphrosyne*** (248 km), ***Iris*** (203 km), ***Hebe*** (192 km), ***Achilles*** (147 km), ***Astraea*** (125 km). There is also a major asteroid beyond 8 times the earth radius, called ***Chiron***, which is 300 km in diameter. More information on the asteroids is given in Ref. [27], Table 3, p. 482.

The destruction of the dinosaurs 65 million years ago was probably caused by the collision with either an asteroid or a comet, about 10 kilometers in diameter. Therefore, we would be seriously harmed if the earth were hit by an asteroid, even one that is very much smaller than the 11 asteroids listed above.

Other Planet Characteristics

Table 4-2 gives additional information concerning the planets. Column (1) shows the ratio of the maximum to the minimum radius of the planet orbit. For most of the planets, this ratio is very close to unity, which indicates that the orbits are nearly circular. Pluto and Mercury have highly elliptical orbits, with ratios of 1.659 and 1.516, respectively. Pluto is a very unusual planet, and so the high ellipticity of the Pluto orbit is not unexpected. The gravitational field of the sun is so strong in the vicinity of Mercury that Mercury's orbit is distorted by relativistic effects. These relativistic effects may also have contributed to the ellipticity of the Mercury orbit.

Column (2) shows the inclination of the planet orbits relative to the ecliptic, which is the plane of the earth's orbit around the sun. Except for

Mercury and Pluto, the inclination angles are very small. This shows that nearly all of the orbits rotate in essentially the same plane. This fact is important in theories to explain how the solar system was formed.

The inclination of the orbit of Pluto is quite large (17 degrees). This again supports the belief that the unusual planet Pluto was a comet that was captured from the Kuiper comet belt.

The inclination of the orbit of Mercury is twice as large as the maximum inclination of the other normal planets. This may be the result of the very high gravitational field of the sun in the region of Mercury.

Table 4-2: Orbit and spin characteristics of planets and spin of the sun.

Planet	Orbit Ellipse Ratio	Inclination of Orbit (deg)	Spin Period	Spin Tilt (deg)
Mercury	1.516	7.0	58.7 d	0.0
Venus	1.014	3.4	243 d	177 R
Earth	1.040	0.0	23.935 h	23.45
Mars	1.210	1.9	24.6 h	25.2
Jupiter	1.103	1.3	9.9 h	3.1
Saturn	1.114	2.5	10.7 h	26.7
Uranus	1.096	0.8	15.6 h	97.9 R
Neptune	1.020	1.8	18.4 h	29.6
Pluto	1.659	17.1	6.4 d	118 R
Sun			27.3 d	0
	(1)	(2)	(3)	(4)

Column (3) gives the spin periods of the planets, along with the spin period of the sun. The earth spins with a period of 23.975 hours. This is the time for the earth to rotate one complete rotation relative to the stars. The earth takes exactly 24 hours to rotate one complete revolution relative to the sun. The earth spins faster with respect to the stars, than with respect to the sun, because the earth revolves around the sun as it spins on its axis. Every year, the earth spins one extra turn with respect to the stars than it does with respect to the sun.

Column (4) shows the tilt of the planet spin axis relative to a perpendicular to the plane of the orbit. The letter *R* indicates that Venus, Uranus, and Pluto spin in a ***retrograde*** direction, opposite to the motion of the planet in its orbit. For the earth, the tilt angle of the spin axis is 23.45 deg. This tilt of the earth axis is what establishes our seasons as the earth rotates around the sun. During summer in the northern hemisphere, the North Pole is tilted toward the sun, and during the winter it is tilted away from the sun.

Venus has a spin period of 243 days, and so it is barely moving. This is greater than the orbital period (225 days) shown in Table 4-1. The tilt of the spin axis in column (5) is 177 deg (with the letter R representing *retrograde* rotation). This is (180 - 3) degrees, and so is similar to a 3-degree tilt angle of the spin axis. Venus spins in a **retrograde** manner; its spin is opposite to the motion of Venus in its orbit. Relative to the sun, in one Venus year (225 days) Venus spins (1 + 225/243) = 1.93 revolutions. Hence Venus spins one revolution relative to the sun every 117 earth days. With this very slow spin relative to the sun, temperatures are not equalized, and this probably contributed to the extreme global warming on Venus.

Jupiter has a spin period of 9.9 hours, and so spins 2.4 times faster than the earth. Since the radius of Jupiter is 11.9 times the earth radius, the velocity of a point on the Jupiter equator exceeds that of the earth by the factor (2.4x11.9), which is 29. Thus, points on the Jupiter equator are moving 29 times faster than points on the earth equator.

The spin period of the sun is 27.3 days. This is an average period; the sun spins faster at the equator than at the poles. Since the sun's radius is 109 times the earth's radius, the velocity of a point on the sun's equator is about 4 times the velocity on the earth's equator.

Major Satellites of the Planets

The major satellites of the planets are listed below, along with the planet diameter and the diameter of each satellite. More information on these satellites is given in Ref. [27], Table 2, p. 480.

Mercury (4878 km): none
Venus (12,104 km): none
Earth (12,756 km): moon (3476 km)
Mars (6795 km): Phobos (27 km), Deimos (15 km)
Jupiter (143,000 km): Ganymede (5262 km), Callisto (4800 km), Io
    (3660 km), Europa (3138 km), plus 12+ others below 270 km.
Saturn (120,500 km): Titan (5150 km), Rhea (1530 km), Iapetus
    (1440 km), Dione (1120 km), plus 18+ others below 512 km.
Uranus (51,000 km): Titania (1580 km), Oberon (1524 km), Umbriel
    (1172 km), Ariel (1138 km), plus 11+ others below 472 km.
Neptune (50,500 km): Triton (2700 km), plus 7+ others below 436 km.
Pluto (2320 km): Charon (1186 km).

Jupiter, Saturn, Uranus, and Neptune all have rings, but only Saturn's rings are bright enough to be seen from earth. Jupiter and Saturn each have a satellite (Ganymede and Titan) that is nearly the size of Mars. These satellites might be candidates for life if they were not so cold. They are so far from the sun there is little hope that they have life.

The Comet Belts

Astronomers believe that beyond Pluto there is a belt of 10 million to one billion comets (called the Kuiper belt) that lie close to the plane of the planets. The belt probably contains thousands of comets more than 50 km across. This belt extends beyond Pluto at distances of 40 to 1000 times the radius of the earth's orbit.

There also appears to be a more distant cloud of comets in randomly oriented orbits in the Opik-Oort cloud. This cloud is believed to contain one trillion comets with a total mass equal to the mass of all of the planets. [72] (p. 206) The cloud extends to 3000 times the Neptune orbit radius, or one-third of the distance to the nearest star.

The Kuiper and Opik-Oort clouds of comets have been deduced by studying the orbits of comets that are observed within our solar system. Most comets are like dirty snowballs, about one kilometer in diameter, but some are tens of kilometers in diameter. The tail of the comet is produced by gasses boiled away from the comet body by the heat of the sun. The tail is forced away from the sun by pressure from the sun's radiation, and so the tail always points away from the sun.

Although the tail of a comet has a dramatic appearance, no harm would result if the earth passes through a comet's tail. However, a catastrophe would occur if the body of a comet should hit the earth.

Creation of Our Solar System

Let us consider the process that created our solar system. To help understand how our solar system was formed, we must consider the principle of angular momentum. Figure skaters apply this principle when they execute the maneuver that allows them to spin rapidly.

Principle of Angular Momentum. To execute a spin, a figure skater gives the body a twist by pushing against the ice. The arms are initially stretched out, and then are drawn close to the body. The skates are held in such a way that no force is exerted on the ice as the arms are drawn inward, and so the angular momentum stays constant. This requires that the body spin rapidly when the arms are held close to the body.

4. The Solar System and the Stars   65

The angular momentum principle is illustrated in Fig. 4-1. Two equal masses are connected together, and are rotating about the center of rotation. Each mass is at a radial distance r from the center of rotation. The velocity of each mass is denoted V, as indicated by the two arrows. The mass of each body is defined as M/2, so that the total mass is M. The angular momentum of these two rotating masses is equal to

Angular momentum = MVr

(Note that MVr means M, times V, times r.) If no torque is applied to the rotating masses, the angular momentum stays constant.

These two masses could represent the two arms of the skater. When the arms are drawn toward the body, the radial distance r decreases. Since the angular momentum (MVr) stays constant, the velocity V increases as the radius r decreases. Therefore, bringing the arms close to the body makes the skater spin faster.

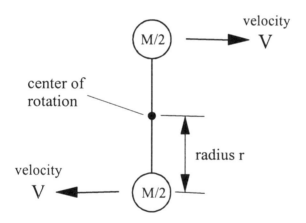

Figure 4-1: Pair of equal masses tied together and rotating about center of gravity; angular momentum is (M V r)

If the radius r is reduced to half, the velocity V must double for the angular momentum to stay constant. The frequency of the rotation, which is the number of revolutions per second, is proportional to the ratio V/r. Therefore, when the radius r is reduced to half, the velocity V is doubled, and the frequency of rotation (V/r) increases by a factor of 4. Consequently, the frequency of rotation becomes very high when the arms of the skater are drawn tightly against the body.

The sun experienced a similar effect when it was formed from a large gas cloud. The diffuse gas cloud was initially rotating because of the rotation of our whole Milky Way galaxy. As the gas cloud coalesced, it would have rotated faster and faster if the angular momentum stayed constant. Eventually, the centrifugal force from the rotation would have kept the cloud from collapsing further. If this had occurred, the sun would never have formed. There must have been a process that allowed the sun to lose its excess angular momentum as the cloud collapsed.

Our sun has 99.87 percent of the total mass of the sun plus its solar system, but has only 2 percent of the angular momentum. Jupiter has 70 percent of the angular momentum, and Saturn has nearly all of the rest. This suggests that as the sun-cloud collapsed, the excess angular momentum was transferred from the sun to a surrounding disk of gas and dust, which eventually formed the planets.

Therefore, a solar system around a star is probably a normal development in the creation of the star. A solar system is needed to allow the star to lose its excess angular momentum and thereby continue to collapse. This conclusion is supported by observations of infrared stellar radiation. These observations appear to show that stars less than 400 million years old are usually surrounded by dust clouds, but older stars are usually not. This suggests that planets formed around the older stars have swept up the dust clouds. [39]

Alfven Theory of Plasma Electric Currents

How was angular momentum transferred from our sun to its solar system? As reported by Lerner [16] (pp. 188-189), Nobel Prize winner Hannes Alfven (1908-1995), who was the pioneer in plasma physics research, explained that the thin gas in space is ionized to form what is called plasma. The electrons are separated from the atoms to create electric currents that flow in space. The current that flows through a square meter of space is very small, but the total current flowing over the vast expanse of space is huge. These electric currents produce magnetic fields that generate enormous torques. Alfven concluded that the magnetic fields produced by plasma electric currents transferred angular momentum from the sun to its surrounding disk of gas and dust. This disk eventually coalesced to form our solar system.

We are therefore led to the following explanation of how our sun and its solar system were created. As the cloud of gas and dust compressed to form the sun, a disk of gas and dust remained around the condensing sun. Plasma electric currents flowed through this ionized gas to produce a large magnetic field. The magnetic field of the central sun rotated more rapidly than that of the surrounding disk. The interaction of these two magnetic fields created a torque that transferred angular momentum from the sun to the surrounding disk, thereby reducing the rotation rate of the sun. The effect is similar to the torque that is produced in an electric motor.

We see the effects of plasma electric currents in space by the northern lights (aurora borealis), which light up the sky when charged ions from the sun hit the upper atmosphere of the earth. Besides generating dramatic lights, these electric currents in space can severely disrupt radio communications. Alfven came from Sweden, where northern lights are often visible. Over many years, he performed extensive experimental and theoretical research on the plasma effects that are responsible for northern lights, and was finally rewarded with the Nobel prize for his pioneering research.

Alfven demonstrated experimentally that plasma electric currents in space experience instabilities, which cause the electric currents to spin around one another like the strands in a rope. From these studies, Alfven concluded that plasma electric currents are fundamentally responsible for all of the spins of bodies in space. He concluded that plasma electric currents in space are the mechanism that caused our whole Milky Way galaxy to spin.

Astronomers have generally rejected the theories of Nobel laureate Hannes Alfven concerning the effects of plasma physics on the development of the universe. However, they have not presented viable alternatives to the Alfven theories, nor have they given responsible reasons for rejecting them. This point will be discussed in Chapter 9.

Possibility of Life outside Earth in our Solar System

As a boy I was delighted by stories of Buck Rogers traveling in his rocket ship to visit the strange populations throughout our solar system. Now we know that high-order life is impossible outside the earth within our solar system. The most that we can reasonably hope for is microscopic life on Mars.

Until we sent space probes to Venus, Venus looked like an ideal candidate for life. Venus was often called our "sister planet" because it is about the same size as the earth, and is reasonably close to our distance from the sun. However, Venus is heavily enshrouded in clouds, and astronomers could not tell what was under the clouds.

When we finally sent space probes to look under the Venus clouds, we found that our beautiful sister planet, the Venus goddess of love, was actually the Hades home of Satan. The temperature on the surface of Venus is so hot it would melt lead.

Why is Venus so hot? Since Venus is 73 percent of the distance of earth from the sun, it receives twice as much energy from the sun. However, this is only a small part of the problem. Carbon dioxide and methane in the atmosphere of Venus have produced a run-away greenhouse effect that produced its hellish environment.

We saw earlier that Venus rotates extremely slowly relative the sun. A Venus day is 117 earth-days long. Because of this very slow rotation, the atmosphere is not churned up with prevailing winds as it is on earth. This effect probably augmented the greenhouse effect.

There are many who pooh-pooh the problems of global warming on earth. We are a long way from the condition where lead melts on our rocks, as it does on Venus, but global warming on earth is real and should be taken seriously.

On the other hand, the people who shout the loudest about global warming often do not understand what is causing it, or what can be done to reduce it. The black smoke emitted from a smoke stack causes pollution but does not produce global warming, and neither does the hole in the ozone layer. Global warming is caused by the colorless gasses, carbon dioxide and methane. Adding pollution controls to power plants may make our air healthier, but does not affect global warming.

Pollution controls can remove soot and trace gasses, like carbon monoxide and sulfur dioxide. However, pollution controls cannot remove carbon dioxide, because carbon dioxide and water are the major combustion components when a fossil fuel is consumed.

To illustrate this point, assume that a ton of coal (a 30-inch cube that is mostly carbon) is burned in pure oxygen. This would produce 3.7 tons of carbon dioxide, having a volume equal to 4 medium-sized houses, each house equivalent to a 30-foot cube. If the coal is burned in air, which is 80 percent nitrogen, the mixture of carbon dioxide and nitrogen fed up the chimney would have the volume of 14 medium-sized houses.

Enormous Distances to the Stars

Since there is no possibility of advanced life outside the earth within our solar system, we must go beyond our solar system to a planet around a star to find advanced extraterrestrial life.

After our manned trip to the moon, the concept of space travel became very popular, and we began to experience through fiction, in television, movies, and books, an abundance of fanciful voyages throughout our universe. It is clear that few people who are delighted by these stories realize how infinitesimal our trip to the moon was when compared with a voyage to even the nearest star.

Distances to stars are commonly measured in light years, which is the distance that light travels in one year. Light travels so fast that to our senses it seems to move instantaneously between points on the earth. With satellite telephone communication, we have direct experience that the speed of light is finite.

Communication satellites are in synchronous orbits, located 26,000 miles from the center of the earth. Consequently it takes 0.15 sec for the communication signal to travel from earth to a satellite, and the round trip takes 0.3 sec. This results in an awkward pause in a satellite telephone conversation. If one speaker starts immediately after he hears the other stop, there is a gap of 0.6 seconds between each comment, because two trips to the satellite are involved.

You can see this effect on television. When a commentator talks to a reporter via satellite, the reporter is very slow in responding to a question from the commentator, because of this time delay.

The tests on Mars with the Rover vehicles in 1997 and 2004 have demonstrated a much more serious problem with the limitation of the speed of light. It takes about 10 minutes for a signal to reach Mars, or 20 minutes for the round trip. Therefore if the Mars Rover vehicle observes an obstacle, the earth-based controller cannot take corrective action until 20 minutes later. Consequently controlling the action of this robot on Mars is very difficult.

It takes only 1.4 seconds to transmit light to the moon, which is the farthest point to which man has traveled in space. Light from the sun takes 8.3 minutes to reach earth, and it takes 4.2 hours to reach Neptune. The time for light to travel from the sun to our outermost planet, Pluto, varies from 4.1 to 6.8 hours, because Pluto has a highly elliptical orbit.

However, these times are puny in comparison to the time for light to come from even the nearest of stars. It takes light 4.2 years to travel from the closest star, Proxima Centauri. More worthwhile stars to visit are its much brighter neighbors, Alpha Centauri A and Alpha Centauri B, which are 4.3 light years away.

The Alpha Centauri A and B stars form a binary pair, which orbit around one another in elliptical 80-year orbits. They are separated by 10 to 30 times the earth-orbit radius. The Alpha Centauri stars are about the size of our sun. If they were alone, they would be excellent candidates for stars with planets like earth that could support life. However, it is doubtful that a stable earth-like orbit can exist about a star in a binary-star pair. Consequently, we probably have to look beyond the Alpha-Centauri stars to find stars with planets that are worth-while visiting.

There are 12 stars within a distance of 10 light years, 28 stars within 12 light years, and 62 stars within 17 light years. [26] These results extrapolate to about 100 stars within 20 light years. In our region of the Milky Way galaxy, the average distance between stars is 7 light years. Stars are packed much more closely in the nucleus at the center of our galaxy.

The nearest star, Proxima Centauri (4.2 light-years) is 100 million times further than the moon (1.4 light-seconds), which is the longest space voyage yet made by man. We have sent unmanned space vehicles to observe the outermost planets of our solar system, and it took 12 years to reach Neptune. The nearest star is 9000 times further than Neptune. *At our present rate of space travel, it would take about 100 thousand years to reach the nearest star.*

Limitation Set by the Speed of Light

Although 100 thousand years would be needed to reach the nearest star with our present space technology, the space-travel optimists continue to believe that it is only a matter of time before our improved technology will make interstellar space travel a common occurrence. In the not too distant future, humans will be hopping from star to star just as was done in the *Star Trek* television series.

Before we are carried away with our optimism, we should understand that there is one limitation that human technology can never overcome. No physical object can travel faster than light. If a star is 10 light years away, it will always take more than 20 years to take a round trip to the star. This is a solidly established principle of the Einstein theory of relativity.

There is a prediction of relativity, called the "Twin Paradox", which might seem to refute this principle. If a space traveler can accelerate and decelerate sufficiently, as he takes a round-trip voyage to a star, he would age more slowly than would his twin who stays here on earth. A space traveler might travel to a star 10 light years away and return in 5 years, in terms of his own age. However, his twin, who stays on earth, would have aged by 20 years, and so would now be 15 years older than his voyager twin. This relativistic effect is called "time dilation".

On the other hand, even if it is ever possible to implement time dilation, and thereby achieve an effect that is similar to traveling faster than light, the time for the trip is not shortened relative to the people who stay on earth. Since the earth-living people must finance the space trip, it is their time that really counts. Relative to the time frames of people staying here on earth, the round trip to a star will always take longer than twice the time for light to reach us from the star.

Distribution of Star Sizes

We saw earlier that the formation of a solar system around a star is probably a normal aspect of stellar development. Consequently there may be many planets orbiting stars within our Milky Way galaxy that are similar to earth and might have advanced life.

Most of the stars are very dim, and do not emit enough energy to support life in planets that may orbit them. To determine which fraction of the stars might be candidates for supporting advanced life, consider the distribution of stars of different sizes shown in Table 4-3. These data were obtained from *Story* [2] (p. 36), Figure 3-1.

Table 4-3: Percent of total stars that lie within particular mass ranges, and their corresponding radiated powers

mass relative to sun	percent of stars	radiated power relative to sun
(1) less than 15 %	40 %	less than 0.05 %
(2) 15 % to 40 %	43 %	0.05 % to 2.6 %
(3) 40 % to 60 %	7.5 %	2.6 % to 13 %
(4) 60 % to 100 %	5 %	13 % to 100 %
(5) 1 to 2	2.5 %	100 % to 16 times
(6) more than 2 times	2 %	greater than 16 times

In Table 4-3, line (1) shows that 40 % of the stars have less than 15 % of the sun's mass and radiate less than 0.05 % of the sun's power. (Stars with less than 1/12 of the sun's mass do not ignite and therefore remain dark bodies.) Line (2) shows that 43 % of the stars have between 15 % and 40 % of the sun's mass, and radiate between 0.05 % and 2.6 % of the sun's power. Thus, lines (1) and (2) show that 83 % of the stars are so dim they radiate less than 2.6 % of the sun's power. These dim stars are highly unlikely to have planets with advanced life.

Line (3) shows that 7.5 % of the stars have 40 % to 60 % of the sun's mass, and radiate between 2.6 % and 13 % of the sun's power. It also seems unlikely that these stars have planets that can support advanced life.

Line (6) shows that 2 % of the stars have more than 2 times the mass of the sun and radiate more than 16 times the sun's power. (The heaviest stars that have been found have 100 times the sun's mass.) These heavy stars in line (6) have very short lives because they burn up their nuclear fuel so rapidly. A star with twice the mass of the sun has 1/8 of the sun's lifetime.

Our sun will spend 10 billion years in the primary phase of its life, while it fuses hydrogen to form helium, and then will enter the twilight phase, which we will discuss later. A star with twice the mass of our sun will spend only 1.25 billion years in the major phase of its life. This time is far to short for advanced life to develop. It took over 4 billion years after our sun was formed before animals appeared on earth.

Thus the stars in line (6) burn up too fast, and the stars in lines (1) to (3) are too dim to support advanced life. This leaves lines (4) and (5) that contain stars within the size range that might support advanced life. The total of these two lines is 7.5 %. Therefore we conclude that in our search for advanced life, no more than 7.5 % of the stars are worthwhile to visit. Probably 5 % is a better number --- one star out of 20.

There are about 100 stars within 20 light years. Probably no more than 5 to 10 percent of these stars are profitable to visit if we are searching for advanced life. This gives us 5 to 10 promising stars within 20 light years. To take a round trip to a star 20 light years away will take at least 40 years regardless of how much our technology may advance. In other words, the *Star-Trek* concept of interstellar travel may make interesting fiction, but has little relation to physical reality.

Measurement of Stellar Distances

As we discuss the enormous distances to the stars, the reader may wonder, "How do we know that the stars are actually that far away?" Let us examine the techniques used to measure stellar distances.

Parallax Method for Measuring the Distance to a Star

The distances of stars within 200 light years can be measured by the parallax method. As the earth revolves around the sun, the images of nearby stars shift over the year relative to the distant stars. This shift is called parallax, a process that our eyes and brain use to give us depth perception.

The images received by our two eyes are not exactly the same, and our brain compares these images to distinguish close objects from distant ones. To observe this effect, hold a finger in front of your nose and look with one eye at a time. As you switch from one eye to the other, the image of the finger moves side-to-side relative to the distant background. The closer the finger is held to the nose, the greater is the relative motion. The reader should stop and perform this experiment in order to understand the parallax principle.

In like manner, the motion of the earth around the sun causes the image of a nearby star to move relative to the distant stars by an amount that is inversely proportional to distance. The motion is ±1.0 arc second for a theoretical distance of 3.26 light-years. (This distance, 3.26 light-years, is called a *"parsec"*, and is the primary distance measurement unit used by astronomers.) If a star shifts by ±0.1 arc seconds over the year, the star is 32.6 light-years away (10 parsecs).

With modern telescopes, astronomers can measure with parallax the distances to stars out to 200 light-years. The measurement is accurate to one light-year for a distance of 30 light-years, but accuracy is very poor for stars at 200 light-years.

There are 100 stars within 20 light-years. The volume of a sphere is proportional to the cube of its radius, and so the number of stars is approximately proportional to the cube of the distance. Consider for example the stars within 40 light-years (2 times 20 light years). Since the cube of 2 is 8, there should be about 8 times as many stars within 40 light-years as there are within 20 light-years. There are 100 stars within 20 light-years, and so there are about 800 stars within 40 light-years.

This reasoning shows that there are about 12,000 stars within 100 light-years, and 100,000 stars within 200 light-years. Since parallax can be used with moderate accuracy out to 100 light-years and with crude accuracy out to 200 light-years, astronomers can use parallax to measure distances to 12,000 stars with moderate accuracy, and the distances to 100,000 stars with crude accuracy.

With modern telescopes, particularly the Hubble space telescope, distance measurements by parallax have improved immensely.

Classifying Stars by Spectra to Estimate Distances

How do we find the distance of a star that is too far away to be measured by parallax? The most fundamental technique for measuring the distances of a remote star is to classify stars in terms of their spectral types. Stars of the same spectral type usually radiate roughly the same light power. The power received from a star is compared with the power received from a close star of the same spectral type. If the distance to the close star is known, one can estimate the distance of the remote star by recognizing that received power varies inversely as the square of the distance. For example, if the light power received from a close star is 100 times the power from a remote star of the same spectral type, the remote star is estimated to be 10 times further away than the close star.

Radiation Spectrum from an Ideal Blackbody

To measure the spectrum of a star, the starlight is fed through a device that acts like a prism, which separates the wavelengths of the light into a rainbow pattern. The relative power at the different wavelengths of the rainbow pattern gives the spectrum of the star.

Astronomers frequently use the spectrum of an *ideal blackbody* in their studies. The radiation from a star can usually be approximated by the spectrum of an *ideal blackbody*, with a blackbody temperature equal to the surface temperature of the star. Figure 4-2 shows the general spectral plot of an *ideal blackbody*.

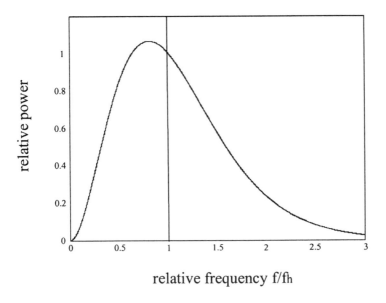

relative frequency f/fh

Figure 4-2: Spectrum of an ideal blackbody radiator

This shows the general spectrum of a blackbody versus frequency. (A plot versus wavelength has a somewhat different shape.) The low frequencies correspond to the red end of the spectrum, and high frequencies correspond to the violet end. The temperature of a blackbody determines the actual frequencies that the body radiates, but the shape of the spectrum is the same for all temperatures.

The frequency scale is expressed in terms of the half-power frequency f_h. Half of the power falls at frequencies less than f_h and half falls at higher frequencies. Area under the curve is proportional to power, and so the areas under the plot on each side of the vertical line are equal. The half-power frequency f_h varies with the temperature of the blackbody.

It is convenient to express the half-power frequency f_h in terms of the equivalent half-power wavelength L_h. The half power wavelength L_h is equal to the ratio c/f_h, which is the speed of light c divided by the half-power frequency f_h.

The half-power wavelength L_h is related as follows to the temperature T at the blackbody, expressed in degrees Kelvin (°K):

L_h = 4.107/T millimeter (mm)

Kelvin temperature is measured above absolute zero temperature, which is -273 degrees Celsius. The temperature in degrees Kelvin is obtained by adding 273 degrees to the temperature in degrees Celsius. At absolute zero temperature, the random motion of molecules is zero.

Let us apply the blackbody spectrum to our sun. From the sun spectrum, one can measure a half-power wavelength L_h of 0.712 micrometers (millionths of a meter), which is equal to 0.000712 millimeters. Assuming an ideal blackbody, the above formula shows that the corresponding blackbody temperature T is 5770 °K (degrees Kelvin). This is approximately the surface temperature of the sun.

Light intensity is defined as the light power P per unit of surface area A, and is the ratio P/A. The intensity of light radiated from an ideal blackbody is related as follows to the temperature T of the body, expressed in degrees Kelvin:

P/A = 5.68 $(T/1000)^4$ watt/cm^2

Our sun approximates a blackbody with a temperature T of 5770 °K. This formula gives an intensity of 6300 watts per square centimeter for the radiation at the surface of the sun.

The light from nearly all stars can be approximated by the radiation from an ideal blackbody. By feeding the starlight through a prism, one can measure the spectrum of the star, which will have the general shape of the ideal blackbody spectral plot of Fig. 4-2. From this spectral recording, one can measure the half-power frequency f_h of the spectrum, or the equivalent half-power wavelength L_h.

If the spectrum is displayed versus frequency, one measures the half power frequency f_h by noting that half of the power falls at frequencies less than f_h, and half falls at greater frequencies. If the spectrum is displayed versus wavelength, one measures the half-power wavelength L_h by noting that half of the power falls at wavelengths less than L_h, and half falls at greater wavelengths. Wavelength L_h is equal to c/f_h.

When the half-power wavelength L_h of the star spectrum has been found, one can calculate from the above two formulas the temperature T of the star surface and the radiation intensity at the star surface.

Physical concept of an ideal blackbody. Let us consider the meaning of a *blackbody* spectrum. An object at room temperature continually radiates energy in terms of heat, and it absorbs heat from the environment. Heat radiation is like light, except that it has a longer wavelength. The blacker the surface of an object, the better the object absorbs radiation, and the better it radiates. A *blackbody* is a physical idealization that radiates the maximum possible energy from a body at any given temperature.

One can build a nearly ideal *blackbody* by machining a spherical cavity inside a block of metal, and cutting a small hole into the cavity. Inside the cavity, radiation is emitted from each portion of the spherical surface, and continually reflects off other surfaces, until a small amount leaks out of the hole. The energy escaping from the hole is close to ideal *blackbody* radiation. A *blackbody* is an idealized physical concept, yet it can be closely approximated by physical equipment.

Application of blackbody spectrum to stars. By approximating a star as an ideal blackbody, one can determine a great deal about the star characteristics. From the general spectral plot of Fig 4-2 one can measure the half-power wavelength L_h of the star spectrum. From this one can calculate with the above equations the blackbody temperature T and the power radiated from the star per unit of surface area (the radiation intensity). If the distance to the star is known, one can determine the power radiated from the star. Dividing the total power by the radiation intensity gives the surface area of the star, and from this one can obtain the star diameter.

Dickinson [28] (pp. 80-81) discusses the use of the spectral characteristics of stars to estimate their stellar characteristics. Temperatures of visible stars vary from 50,000 to 1800 °K, which corresponds to half-power wavelengths from 0.08 to 2.2 micrometers.

Cepheid Variable Stars

An accurate yardstick for astronomy was developed by astronomer Henrietta Leavitt (1868-1921) in 1908, based on Cepheid variable stars. These are stars that vary periodically in brightness, and are named after the star Delta Cephei. Leavitt observed Cepheid variable stars in the Small Magellanic Cloud, which can only be observed in the southern hemisphere. It is a cluster of stars outside our Milky Way galaxy, about 300,000 light-years from earth. All stars in the Small Magellanic Cloud are at approximately the same distance.

The light from a Cepheid variable pulsates, typically varying by two to one in brightness. Leavitt found that the period of brightness variation was precisely related to the light power received from these Cepheid variables. Period is the time for one cycle of brightness variation. Since all stars in the Small Magellanic Cloud are at nearly the same distance, radiated power was approximately proportional to received power.

Astronomers measured some close Cepheid variable stars, the distances of which could be measured by parallax. This allowed astronomers to determine the actual power radiated from a Cepheid variable from its period of variation.

When one finds a Cepheid variable star in a distant galaxy, one can determine from its period of variation the absolute power that the Cepheid variable star radiates. By comparing this with the power received from the star, one can calculate the distance to the Cepheid variable star, by recognizing that received power varies inversely as the square of the distance. This allows one to measure the galaxy distance.

A problem with the early use of the Cepheid variable method of measuring distance is that there are two types of variable stars. This confusion caused a factor of two error in early estimates of galaxy distance.

Life Cycles of the Sun and Stars

Based on the understanding of the nuclear fusion processes being performed within the stars, and on many observations of stars in various stages of their life cycles, astronomers have derived explanations of the life cycles of stars that have strong scientific support. The following is a summary of these concepts.

Our sun has been generating energy by fusing hydrogen to form helium for five billion years, and will continue this process for another five billion years until the hydrogen fuel in the core of the sun is exhausted. Then the sun will contract to achieve a temperature of 100 million degrees Celsius. A second nuclear fusion process will occur in which 3 helium atoms are fused to form one carbon atom. This fusion of helium into carbon will reduce the mass by 0.07 percent, and will release 18,000 kilowatt-hours of energy per gram of helium. This is one-tenth of the energy released per gram when hydrogen is fused to form helium.

When the hydrogen fuel in the sun's core is exhausted, the sun will begin to burn the hydrogen in its outer shell. This will cause the sun to swell enormously to become a red giant star. The surface of the sun will swell at least to half the radius of the earth orbit, and the earth will be burned to a crisp.

The burning of hydrogen in the outer shell of the sun will continue as the core burns helium. When the hydrogen in the shell is exhausted, helium will begin to burn in the shell. This will make the red giant sun unstable, and it will blow off a large amount of material to form what is called a "planetary nebula". About one-third of the sun may be expelled in the hot gasses of this nebula. [43]

The term "planetary nebula" was coined by the famous 18th century English astronomer, William Herschel. He thought that a planetary nebula represented the beginning of a solar system around a star. His guess was wrong, but the name stuck. Planetary nebulae form the most beautiful displays visible in astronomical telescopes. Excellent pictures of these nebulae are given in Ref. [43].

It will take 100 thousand to one million years for our sun to blow off its planetary nebula. The core that remains will have little hydrogen or helium fuel left. Nuclear fusion will cease, and our sun will shrink to become a white dwarf star.

The White Dwarf Star

Heat generated by nuclear fusion produces pressure that offsets the compressive force produced by gravity. When the helium fuel is exhausted, nuclear fusion will permanently stop. With nothing to offset the gravitational pressure, the sun will begin to shrink. As the sun shrinks, gravitational energy is released, which generates heat and pressure to offset gravity. This heat cannot be sustained, and so the sun shrinks steadily, glowing white hot from the release of gravitational energy. The sun will shrink to become a very small white star, which is called a ***white dwarf***. (A similar release of gravitational energy was discussed on page 4, when a hydrogen cloud collapsed to form our sun.)

When our sun reaches the size of the earth, it will stop shrinking and so will stop radiating energy. It will gradually fade to become a black dwarf star, the dead ember of a once brilliant star. It will have a density of about one ton per cubic centimeter, which is one million times the density of water.

One would not want to land a space ship on a black dwarf star. The force of gravity would be one million times that on earth, and the space ship with its inhabitants would be crushed.

There is abundant evidence that white dwarf stars (with their extremely high densities) actually exist. The brightest star in the sky is Sirius, which has a companion star that is a white dwarf. This white dwarf has been studied extensively. Its mass has been determined from its gravitational effect on the motion of Sirius. Sirius and its white-dwarf companion are 8.6 light-years away.

The basic life cycle that we have discussed for our sun is followed by all stars with less than 8 times our sun's mass. [43] (p. 52) However, the rate at which a star proceeds through its life cycle varies greatly with its mass. A star with twice our sun's mass generates 16 times as much power, and has 1/8 of the lifetime of our sun. A star with one-half of the sun's mass generates 1/16 as much power, and has 8 times the lifetime of our sun. The mass of a star can vary from 1/12 of the sun's mass to 100 times its mass. [27] (p. 250)

The Supernova

A star with more than 8 times our sun's mass ends its life in a radically different manner. It dies in a dramatic explosion called a supernova. A massive star begins its life by fusing hydrogen to form helium, and then fuses the helium to form carbon, just as with smaller stars. However, with its greater mass, the star can achieve the temperature and pressure to generate the following elements by nuclear fusion: oxygen, neon, silicon, nickel, cobalt, and iron. Then the fusion stops, because the generation of elements heavier than iron does not release energy; it absorbs energy. The fusion processes from carbon to iron reduce the stellar mass by another 0.12 percent, and release 30,000 kilowatt-hours of energy per gram of carbon.

When this massive star runs out of nuclear fuel, catastrophic gravitational collapse suddenly occurs. This collapse releases so much energy that the star explodes in a brilliant supernova, which shines with the brightness of billions of suns for a month.

In the supernova explosion, neutrons and neutrinos are fired into the outer portion of the star, creating nuclear reactions that produce elements that are heavier than iron. These heavy elements, along with other elements created earlier in the star, are scattered as dust particles throughout the galaxy by the supernova explosion. These dust particles are picked up in the hydrogen clouds that form new stars, and produce the solid material from which planets like our earth are created.

Only 0.3 percent of the stars have 8 times the mass of our sun or greater, and so will become supernovas. However, 14 percent of the total stellar mass is contained in these massive stars. Consequently, considerable material is spread across the galaxy in supernova explosions. (See *Story* [2], Figs 3-1, 3-2)

A supernova explodes within our Milky Way galaxy about once a century. This represents the death of all stars with masses greater than 8 times the mass of our sun. Planetary nebulae appear in our galaxy about once a year, and represent the later stage of all stars with less than 8 times the mass of our sun. There are probably many more planetary nebulae that are obscured by dust.

The Structure of the Atom

When the sun ends its life, gravitational pressure will squeeze the sun into a state of extreme density. One cubic centimeter of the sun will weigh 2 tons. To understand how this extreme density can occur, let us briefly examine the structure of the atom. A more detailed discussion of the atom is given in Appendix B.

All matter consists of combinations of about 100 different kinds of atoms. An atom is made up of three elementary particles: the electron, the proton, and the neutron. The electron has a negative electrical charge. The proton has 1840 times the mass of the electron, and has a positive electrical charge, equal and opposite to that of the electron. The neutron has no electrical charge and has approximately the mass of the proton. All three elementary particles have been observed as separate particles outside the atom.

The protons and neutrons of an atom are contained within a very compact nucleus. The electrons orbit around the nucleus, forming a spherical electron cloud that typically has about 50,000 times the diameter of the nucleus.

The simplest atom is hydrogen, which has one electron orbiting a single proton. Helium has two electrons, which orbit a nucleus containing two protons and two neutrons.

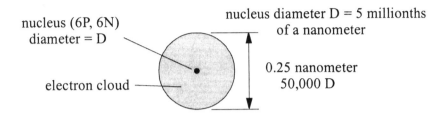

[a] Carbon atom at normal pressure
Density = 2 grams per cubic centimeter

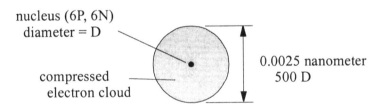

[b] Compressed to white dwarf star
Density = 2 tons per cubic centimeter

[c] Compressed to neutron star
Density = 250 million tons per cubic centimeter

Figure 4-3: Approximate dimensions of a carbon atom under conditions of [a] normal pressure, [b] white dwarf star, and [c] neutron star

Density of the White Dwarf Star

Figure 4-3 illustrates the effect on the atom as a star is compressed to the density of a white dwarf star, and then to a neutron star. Before running out of nuclear fuel, our sun will consist primarily of carbon atoms. Diagram (a) shows the approximate dimensions of a carbon atom at normal pressure. A carbon atom has 6 protons (P) and 6 neutrons (N). The diameter of the carbon atom is about 0.25 nanometer, where a nanometer is one millionth of a millimeter. (The wavelength of visible light is about 600 nanometers.) The diameter D of the carbon-atom nucleus is about five millionths of a nanometer or five trillionths of a millimeter.

Hence the diameter of the carbon atom (0.25 nanometer) is 50,000 times greater than the diameter of its nucleus (5 millionths of a nanometer). The volume of the carbon atom exceeds that of the nucleus by the cube of 50,000. This means 50,000, times 50,000, times 50,000, which is 125 trillion (125 million, million). As shown, the density of carbon is about 2 grams per cubic centimeter.

Since the electrons and protons have opposite electrical charges, the electrons are attracted to the protons contained in the nucleus. However, the electrons do not fall into the nucleus, because they are rapidly orbiting around the nucleus. The orbiting of the electrons around the nucleus is similar to the planets orbiting around the sun. However the planets travel in nearly a single plane, whereas the electron orbits fill the three-dimensional space around the nucleus. The electrons form a three-dimensional cloud that surrounds the nucleus.

When our sun runs out of nuclear fuel and shrinks to a white dwarf star, its diameter will decrease by 100, and so its density will increase by the cube of 100, which is one million.

Diagram (b) in Fig. 4-3 shows what happens to the carbon atom in diagram (a) when it is compressed to the ultimate density of a white dwarf star. The diameter of the carbon atom is reduced from 0.25 nanometer to 0.0025 nanometer. Density is increased by one million, from 2 grams per cubic centimeter in (a) to 2 metric tons per cubic centimeter in (b). A metric ton is 1000 kilograms or one million grams, and is 10 percent greater than an English ton (2000 pounds).

Diagram (b) shows the white-dwarf state, which is the final condition for stars with masses less than 8 times that of our sun. A density of 2 tons per cubic centimeter is enormous, yet the atom still has a large amount of space left in the electron cloud circling the nucleus. The force from the electrons is sufficient to resist the enormous gravitational pressure of the very compact white dwarf star.

The ***Pauli Exclusion Principle*** is an important concept of quantum mechanics, which states that electrons can only occupy specific orbits (or more precisely, specific "energy states"). When an electron cloud has been compressed to the final white-dwarf condition, there are no spare energy states left for the electrons to occupy, and so the electron cloud cannot shrink any further.

Density of the Neutron Star

As was stated earlier, when a star with more than 8 times the mass of our sun runs out of nuclear fuel, catastrophic gravitational collapse suddenly occurs. An enormous amount of gravitational energy is released, to produce a supernova explosion.

In this process, the electron cloud can no longer withstand the force of gravity. The gravitational pressure is so great that the electron cloud is destroyed. The electrons are forced into the protons, converting them into neutrons. Since the electrons and protons are eliminated, the star consists entirely of neutrons. This star is called a ***neutron star***.

With the electron cloud eliminated, the atom collapses to the size of its nucleus, as shown in diagram (c) of Fig. 4-3. Relative to a white dwarf, the diameter of the atom decreases by a factor of 500 and so its density increases by the cube of 500, which is 125 million. The density increases from that of a ***white dwarf*** (2 tons per cubic centimeter) to that of a ***neutron star*** (250 million tons per cubic centimeter). (A more accurate value for neutron star density is 200 million metric tons per cubic centimeter, or approximately one billion tons per teaspoon.)

To achieve a physical feeling for the tremendous density of a neutron star, consider a fictitious exercise. With earth-moving equipment, dig a hole 1500 ft deep over an area of 50 acres, to cover a square plot 1500 ft on a side. This material is put into a super-compactor, which compresses all of the material into a volume of one cubic centimeter. That cube would have the density of a neutron star, about 200 million tons per cubic centimeter.

Although the density of a neutron star may seem unbelievably high, it is certainly not infinite. It is the same density that exists here on earth within the nucleus of every atom. The electrons of the atom are eliminated in a neutron star, and so the density of the star increases to the density of the atomic nucleus.

A neutron star consists entirely of tightly packed neutrons. Since there is no space left for further contraction, a neutron star has reached the greatest possible density of matter, one billion tons per teaspoon.

The Pulsar

The neutron star was originally only a theoretical prediction until the pulsar was discovered in 1968. A pulsar emits a radio signal consisting of pulses at precisely timed intervals. When first observed, some people thought that these pulses might be radio signals from intelligent beings on a distant planet. However, studies soon indicated that they are probably generated by rapidly spinning neutron stars.

A pulsar has been found at the location of a supernova that exploded in 1054 AD, and was recorded by Chinese astronomers. This pulsar is surrounded by the Crab Nebula, which is a cloud of gas produced by the supernova. This pulsar generates precisely timed radio pulses at a rate of 30 times per second. It is slowing down at a rate that suggests an origin about 900 years ago. The fastest known pulsar generates nearly 1000 pulses per second. (See Silk [21], p. 325.)

The theory indicates that a pulsar is a rapidly spinning neutron star. The magnetic field of the neutron star radiates radio beams from north and south magnetic poles along the magnetic axis. Like the earth, the magnetic axis does not coincide with the spin axis, and so the radio beams spin in space as the neutron star rotates. When a radio beam passes the earth, we receive a pulse.

There is no nuclear process occurring within a neutron star to generate energy. The energy that is creating the radio beam of the pulsar is apparently derived from atoms that are being sucked onto the neutron star.

The radio beam from a pulsar is narrow. Consequently, we can receive pulsar signals only if the pulsar is oriented so that its beam passes the earth. There are probably many pulsars within our galaxy that are not oriented ideally, and so cannot be seen.

The pulsar star rotates at the pulse frequency. A pulsar with a frequency of 30 pulses per second is a star rotating at 30 revolutions per second. For a star to spin at the frequency of a pulsar, it must be extremely compact. The only explanation that satisfies the pulsar characteristics is a neutron star. Therefore, even though a neutron star has very strange properties, there is strong evidence that neutron stars actually exist.

Chapter 5

Hubble Expansion of the Universe

What Is a Nebula?

Up until the early 1900's, most astronomers believed that our Milky Way galaxy comprised our complete universe. Stars look like points of light to the telescope, but our heavens also contain fuzzy extended objects that were called nebulae. Certain astronomers believed that some nebulae were distant galaxies, like our own Milky Way galaxy, but this concept was widely disputed. The general attitude is illustrated in the following excerpt from a popular 1905 book on astronomy by Agnes Clerke, entitled *The System of the Stars* [19] (p.1):

"The question of whether nebulae are external galaxies hardly any longer needs discussion. It has been answered by the progress of research. No competent thinker, with the whole of the available evidence before him, can now, it is safe to say, maintain any singular nebula to be a star system of co-ordinate rank with the Milky Way. A practical certainty has been attained that the entire contents, stellar and nebula, of the [celestial] sphere belong to one mighty aggregation, and stand in ordered mutual relations within the limits of one all-embracing scheme."

In this quotation, Clerke was insisting that all nebulae lie within our Milky Way galaxy.

The mysterious nebulae had been observed for many years. In 1784, the locations of the larger nebulae were listed by Charles Messier (1730-1817) in his *Catalogue of Nebulous Objects*. This catalogue became the primary reference for locating nebulae, and still serves that purpose today. The name M51 for the Whirlpool galaxy, shown in Fig 1-1 and on the front cover, is the designation given by Messier.

86   *Beliefs and Facts of Creation*

Messier was searching for comets, and he prepared his nebulae catalogue to distinguish comets (which move) from nebulae (which are fixed). Although Messier discovered 15 comets, it is his *Catalogue of Nebulous Objects* that made the Messier name famous. Messier lost his astronomer job when the French Revolution erupted in 1789.

After Clerke's book, astronomers developed methods of measuring stellar distances, and were able to prove that some nebulae lie outside our Milky Way galaxy. Certain nebulae are stellar clusters and gaseous clouds, which lie within our Milky Way galaxy, but others (such as M51) are distant galaxies that are far beyond our Milky Way galaxy.

The reason for this confusion is that it is extremely difficult to measure the distances of stars. Shortly after Clerke's book was published, reliable means for estimating stellar distances were achieved, and this produced a revolution in astronomy.

Measuring the Radial Velocity of a Star

Although it is extremely difficult to measure the distance to a star, the velocity of a star in the radial direction can be measured easily and accurately. Radial velocity is the component of velocity either toward us or away from us, along a "radial" line from earth to the star.

An astronomer studies the light from a star by passing the light through an instrument that acts like a prism, to separate the wavelengths into a rainbow pattern. From this spectral rainbow pattern an astronomer can accurately measure the radial velocity of the star. It is extremely difficult to measure velocity in the transverse direction, perpendicular to the radial line to the star.

An element in the atmosphere of a star generates a unique pattern of spectral lines. These consist of bright lines that occur at wavelengths where an atom radiates energy, and dark lines that occur where an atom absorbs energy. Observations of spectral lines have shown that by far the most abundant elements of the universe are hydrogen and helium. In terms of mass, the cosmic concentrations of the ten most abundant elements are: hydrogen (74.8 %), helium (23.8 %), oxygen (0.88 %), carbon (0.40 %), neon (0.19 %), iron (0.13 %), nitrogen (0.10 %), silicon (0.08 %), magnesium (0.07 %), sulfur (0.05 %). [72] (p. 31)

In the light from a star, the spectral lines are shifted in wavelength by an amount approximately proportional to the radial velocity between the star and the earth. This wavelength shift is called the Doppler effect, named after the Austrian scientist, Christian Doppler (1803-1853), who discovered it in 1842. If the star is moving away from us, the spectral lines are shifted toward the red end of the spectrum, and so the star is said to have *redshift*. If the star is moving toward us, the spectral lines are shifted toward the blue, and the star has *blueshift*.

The Doppler wavelength shift, divided by the normal wavelength of a spectral line, is approximately equal to the radial velocity of the star divided by the speed of light. The exact formula for Doppler shift derived by Einstein is given in *Story* [2], Appendix E, page 236.

Hubble's Discovery of the Universe Expansion

One of the first astronomers to prove that some nebulae are distant galaxies was Edwin Hubble. He had the advantage of working on the new Mount Wilson telescope, which was the best telescope in the world at that time.

In Chapter 4 (p. 77) we discussed the use of Cepheid variable stars to measure the distances of stars. Hubble was able to resolve Cepheid variable stars in the galaxy M31 of the constellation Andromeda, which is 2.3 million light-years away, and in the M33 galaxy of the constellation Triangulum, which is 2.6 million light-years away. This allowed him to estimate the distances to M31 and M33. (The names M31 and M33 are the designations given by Messier in his 1784 *Table of Nebulous Objects*.)

About 15 years later, astronomers discovered that there are two types of variable stars: the classical Cepheid variable stars and the RR Lyrae stars. [19] (p. 35) This confused Hubble's measurement, and so he estimated the M31 and M33 distances to be half of the actual distances.

Hubble examined the brightest stars in the M31 and M33 galaxies, and assumed these to be super-giant stars. He assumed that the brightest stars in more distant galaxies would radiate the same power as those in the M31 and M33 galaxies. With this approach he estimated distances to several galaxies more distant than M31 and M33, which were too far away to observe Cepheid variables.

Hubble categorized these galaxies into different types depending on their shapes. He assumed that all galaxies of a given shape have approximately the same diameters in light years. With this principle, he estimated the distances to galaxies much further away.

Hubble compared his distance measurements with the radial velocities of the galaxies determined from the Doppler wavelength shifts obtained from the galaxy spectra. He discovered that, except for the close galaxies, M31 and M33, all galaxies have redshift and so are moving away from us. He found that the radial velocity of a galaxy is approximately proportional to its distance. This discovery was published by Hubble in 1929, and created a revolution in astronomy. It showed that our universe is expanding. This expansion of the universe is called the Hubble expansion, and the ratio of galaxy velocity divided by galaxy distance is called the *Hubble constant*.

There were many sources of error in Hubble's measurements of galaxy distance. His galaxy distance estimates were 8 times too short and so his Hubble constant was 8 times too large. One source of error was the confusion between the two classes of variable stars, as discussed previously. Another problem was that some of the "super-giant" stars that Hubble observed were actually regions of brightly illuminated hydrogen gas. [19] (p. 35)

Nevertheless, even though Hubble's initial measurement of the Hubble constant had serious errors, his basic discovery was correct. Our universe is indeed expanding.

Meaning of the Hubble Expansion

The meaning of the Hubble expansion of the universe can be clarified by considering a rubber band that is stretched at a constant rate. At an instant of time (called t_1) mark a series of dots on the band separated from one another by 10 mm, and label these dots as follows:

```
D'   C'   B'   A    B    C    D    E
*    *    *    *    *    *    *    *
```

The distance from A to B is 10 mm, from A to C is 20 mm, from A to D is 30 mm, etc. Now look at the band later (at time t_2) when it has stretched by 10 percent, so that the distance between neighboring dots has increased to 11 mm. The distance from A to B is now 11 mm; from A to C is 22 mm; from A to D is 33 mm, etc. Between time t_1 and time t_2, point B moves 1 mm away from A, point C moves 2 mm away from A, point D moves 3 mm away from A, etc. The relative velocity between points along the band is proportional to the distance between the points. This indicates that the rubber band is being stretched at a constant rate.

We can extend this analogy to two dimensions by considering a balloon being blown up at a constant rate. Mark on the surface of the balloon an array of dots. As the balloon expands, the dots separate from one another at velocities proportional to the distances between the dots.

Modern Measurements of the Hubble Constant

Modern telescopes (particularly the Hubble Space Telescope) have allowed astronomers to make accurate measurements of the Hubble constant. Cepheid variable stars can now be observed in very distant galaxies. Another approach uses a particular type of supernova (Type 1a), which radiates a nearly constant amount of power. These supernovas radiate a peak power of about three billion suns for a few weeks, and so can be observed in very distant galaxies.

A Type 1a supernova is apparently formed when a white dwarf star closely orbits a red giant star, and sucks off material from the red giant star. When the mass of the white dwarf reaches a critical value, the white dwarf suffers catastrophic gravitational collapse and explodes as a Type 1a supernova.

Two teams have recently measured the Hubble constant accurately; one led by Alan Sandage and the other by Wendy Freedman. The Sandage study was based on supernovas and the Freedman study was based on Cepheid variables. The average Hubble constants recently reported by Sandage and Freedman are 18.7 and 21.5 km/sec per million light-years. *This book assumes a Hubble constant of 20 km/sec per million light-years, which is a good average of these recent measurements.*

The Hubble constant is usually expressed in terms of the parsec, which is 3.26 light-years. The Sandage and Friedman data are usually specified as 61 and 70 km/sec per megaparsec (million parsecs). [36]

The Observable Universe

Let us assume that the universe is expanding uniformly at our assumed Hubble constant of 20 km/sec per million light-years. If a galaxy is 10 million light-years away, it should recede (move away from us) at a velocity of 200 km/sec. A galaxy 100 million light-years away should recede at 2000 km/sec, and a galaxy 1000 million (or one billion) light-years away should recede at 20,000 km/sec. Therefore, a galaxy that is 15,000 million (or 15 billion) light-years away should recede at 300,000 km/sec, which is the speed of light.

If a galaxy is moving away from us at the speed of light, it cannot be seen. This argument indicates that 15 billion light-years should be our limit of observation. We presumably cannot observe a galaxy more distant than 15 billion light-years. Consequently our *observable universe* is generally considered to be a sphere with a radius of 15 billion light-years. (How can a galaxy travel faster than the speed of light? This is one of many confusing concepts of the Big Bang theory.)

The Apparent Age of the Universe

An obvious explanation of the Hubble expansion of the universe is that the universe began billions of years ago as a highly dense mass that exploded with a Big Bang and has been expanding ever since. Let us assume that the universe has always expanded at the same rate, 20 km/sec per million light-years. With this assumption we can easily extrapolate the universe backward in time. Since a galaxy 15 billion light-years away would recede at the speed of light, our universe should have begun as a very dense mass 15 billion years ago. This book calls this time period, 15 billion years, the **apparent age of the universe**, which is denoted T_0.

There are many versions of the Big Bang theory, but essentially all of them assume that the Hubble constant was greater in the past. Consequently, they compute a true age of the universe that is less than the apparent age. Most versions of the Big Bang theory predict a true age of the universe that lies between $(2/3)T_0$ and T_0, or between 10 and 15 billion years.

Theories to Explain the Hubble Expansion

There are three major classes of theories that have been proposed to explain the Hubble expansion of the universe. These are: (1) the *Big Bang* theory, (2) the *Steady State* theory, and (3) the theory that the Hubble expansion is an *Apparent* effect, that the universe is not actually expanding. Let us examine these three concepts, starting with theory (3).

The Hubble Expansion Is Apparent

Some scientists argue that the universe expansion is *Apparent*, that the universe is not actually expanding. Effects other than velocity can cause a spectral redshift. These effects may make the universe appear to expand, even though it is not.

Paul Marmet [37] has proposed a redshift effect that has a solid scientific foundation. His theory is explained in *Story* [2], Appendix A.

5. Hubble Expansion of the Universe 91

Marmet has shown that a diffuse cloud of hydrogen gas can produce a redshift.

The Marmet redshift effect cannot be observed in the earth's atmosphere, because the density of gas is too high. It cannot be observed in the laboratory, because it requires too long a path to create a measurable effect. However, Marmet has shown that his effect explains the variation of the redshift of radiation across the disk of the sun. The redshift of radiation from the limb of the sun is greater than that from the center. Thus Marmet has experimental evidence to support his theory.

The Marmet redshift would produce an apparent expansion of the universe that is consistent with the Hubble constant if the universe has an average gas density of 30,000 hydrogen atoms per cubic meter. Many gaseous nebulae have hydrogen densities of 100 billion atoms per cubic meter. However Appendix C of this book shows that estimates of the density of matter in the universe give an average density of only 2 hydrogen atoms per cubic meter. This indicates that the Marmet redshift effect is probably not sufficient to explain the Hubble expansion.

On the other hand, Chapter 9 shows that the Marmet redshift effect gives a promising explanation for the extreme redshift of the quasar.

The Steady State Theory

There is another well known alternative to the Big Bang theory, called the *Steady State* theory. This theory was proposed in 1948 by the noted astrophysicist Fred Hoyle (1915-2001), and supported by astrophysicists Hermann Bondi and Thomas Gold. The Steady State theory was seriously considered by many astronomers until it was eclipsed in the stampede toward the Big Bang theory that occurred in the late 1960's.

The Steady State theory assumes that the age of the universe is infinite. The theory postulates that diffuse matter is being created throughout space to compensate for the Hubble expansion. The required creation of matter is far too small to be observed directly. To keep the density of matter constant as the universe expands, only one hydrogen atom need be created every year within a volume of one cubic kilometer.

The Steady State theory postulates that the diffuse matter gathers into huge clouds that coalesce to form new stars and galaxies. In this manner, our universe continually changes, and thereby stays eternally young even though it is infinitely old.

What could produce this spontaneous creation of matter? The Hoyle Steady State theory did not have an answer. Is the matter created *out of nothing*? This may seem unacceptable, but should be no less acceptable than the Big Bang postulate that the whole universe was instantaneously created *out of nothing* at the Big Bang.

During World War II, George Gamow worked in developing the atomic nuclear bomb in the United States, and Fred Hoyle was engaged in British radar development. After the war, both scientists directed their efforts to astronomy, and wrote popular books on the subject. Gamow explained the universe expansion by postulating that our universe began billions of years ago in an enormous explosion. [13] Hoyle presented his alternate Steady State theory. As a criticism of the Gamow concept, Hoyle used the term *"Big Bang"* to characterize the enormous explosion that presumably created our universe. The name stuck, and since that time the Gamow theory was called the *"Big Bang theory"*.

Up until 1965, both the Big Bang theory and the Steady State theory were seriously considered by astronomers. However, in the late 1960's there was a stampede to the Big Bang theory. The co-sponsors of the Steady State theory, Bondi and Gold, lost interest in what had become an unpopular theory, and in time Hoyle abandoned his theory. [19]

The Hoyle Steady State theory was based on analyses using the Einstein general theory of relativity. This book will describe a new version of the Steady State theory that is based on the Yilmaz theory of gravity. This new version of the Steady State theory does not have the limitations of the original Hoyle theory.

The Big Bang Theory

The Big Bang theory proposed by George Gamow in 1947 was radically different from the *Modern Big Bang theory*, which evolved in the late 1960's. Gamow postulated that at the instant of the Big Bang the initial universe had the density of nuclear matter, which weighs one billion tons per teaspoon. Gamow considered the atomic nucleus to have the greatest possible density of matter. This is the same as the density of matter in a neutron star.

The Modern Big Bang theory was derived from studies of the Einstein general theory of relativity. These studies have convinced theorists that the universe began as a *"singularity"*, which ideally had an infinite density of matter. There are many variations of the Modern Big Bang theory, but all of them predict a mass density that is many, many times greater than the density of the atomic nucleus, which Gamow considered to have the greatest possible density of matter.

This chapter gives detailed discussions of the Gamow Big Bang theory and the Modern Big Bang theory.

The Big Bang Age Dilemma

A serious problem with all versions of the Big Bang theory is that the theory does not allow enough time to explain the development of the universe in a satisfactory manner.

By studying the processes of stellar evolution, astronomers are able to calculate the ages of stars. *Science News* (Sept. 18, 2004, p. 189) reports in the article, "Beryllium data confirm stars' age" that two of the oldest stars of our Milky Way galaxy (within the globular cluster NGC 6397) are at least 13.4 billion years old. The article concludes with, "Various evidence shows that the cluster NGC 6397 is 13.4 billion years old. With other observations pegging the universe age at 13.7 billion years, the first stars must have formed less that 200 million years after the Big Bang".

We saw earlier that the *apparent age of the universe* (derived from the Hubble constant) is only 15 billion years. The *apparent age* represents an upper bound to the *true age* of the universe. Most versions of the Big Bang theory give a true age of the universe that is appreciably less than 15 billion years. Even if we make the optimistic assumption that the true age of the universe is 15 billion years, this assumption allows only 1.6 billion years after the Big Bang to create the oldest stars, which are at least 13.4 billion years old. The above *Science News* quotation reports that cosmologists now generally conclude that the Big Bang occurred 13.7 billion years ago.

The Big-Bang age dilemma is actually much worse than this discussion suggests. As explained by Eric Lerner [16] (pp. 15-32), recent astronomical studies have shown that the universe is not at all uniform. In 1986, Brent Tulley, a University of Hawaii astronomer, (assisted by J. R. Fischer) found that almost all galaxies within 1.5 billion light years are concentrated into huge ribbons, typically one billion light years long, 300 million light years wide, and 100 million light years thick. These ribbons contain curling filaments, a few million light years thick, extending for hundreds of millions of light years.

This study evolved from a mapping of individual galaxies out to 100 million light years. In making this map, it was assumed that the redshift of a galaxy specifies its distance. From this map Tulley and Fischer found (with about 20 exceptions) that all of the thousands of galaxies are concentrated into filaments, a few million light years across. These filaments extend for hundreds of millions of light years, beyond the limits of the map. While performing a later mapping study, astronomer Margaret Haynes concluded after examining the curling galactic filaments, *"The universe is just a bowl of spaghetti."*

To expand his study to 1.5 billion light years, Tulley mapped clusters of galaxies, because there are millions of individual galaxies within that range, too numerous to be mapped individually. From this cluster map Tulley discovered his huge ribbons.

About 1990, Tulley's findings were confirmed by several astronomy teams. The most dramatic is by Margaret J. Geller and John P. Huchra of the Harvard Smithsonian Center for Astrophysics, who are mapping individual galaxies out to 600 million light years, about 200 times as many as in the Tulley-Fischer map of individual galaxies. In their preliminary results, they displayed what they call the "Great Wall", a huge ribbon of galaxies stretching across the region mapped, a distance of 700 million light years. This ribbon, which is 200 million light years wide and 20 million light years thick, corresponds closely to a ribbon mapped by Tulley using clusters of galaxies. Galaxy density inside the ribbon is 25 times greater than outside.

These results seriously contradict the Big Bang theory in two ways: (1) it would probably take at least 150 billion years to form these gigantic structures; and (2) the Big Bang theory predicts a very uniform universe, not the spaghetti-like and ribbon-like structures that are observed.

Lerner [16] (p. 23) explains the universe age problem. Except for the general Hubble expansion of the universe, the maximum local velocity of all galaxies is less than 1000 km/sec, which is 1/300 of the speed of light. Since the time of the postulated Big Bang, a galaxy could move only a distance of 15 billion light years divided by 300, which is 50 million light years. However, if the universe were uniform after the Big Bang, galaxies would have had to move 270 million light years to form the huge ribbons. The age discrepancy is worse than these numbers suggest, because time must be allowed for a galaxy to accelerate and decelerate.

The Gamow Big Bang Theory

As explained earlier, Gamow postulated that the initial universe at the instant of the Big Bang had the density of matter in the atomic nucleus, which is one billion tons per teaspoon. Gamow considered this to represent the greatest possible density of matter. [13]

Gamow estimated the initial size of the universe, but he based this on the portion of the universe that could be seen by the telescope at the Mt. Wilson Observatory. The estimated limit of this telescope at that time was 500 million light years. Gamow chose this limit because he knew that the universe extended at least that far. According to Gamow's calculation, the initial size of the portion of the universe observable by the Mt. Wilson telescope was a sphere with eight times the diameter of the sun. Since the sun's diameter is 1.4 million kilometers, this gives an initial universe diameter of 11.2 million kilometers.

Let us extend Gamow's estimate to include the total observable universe. The observable universe has a radius of 15 billion light years, which is 30 times the 500-million light-year estimated range of the Mt. Wilson telescope. Multiplying the 11.2 million kilometer diameter of the universe (calculated by Gamow) by 30 gives a total diameter of 336 million kilometers for the initial size of the observable universe. By comparison, the orbit of the earth around the sun has a diameter of 300 million kilometers.

We have much better astronomical data available today to apply the Gamow postulate. Appendix C (Eq. C-4) shows that the luminous stars shining within our observable universe (15 billion light years in radius) are equivalent to about 60 billion times one billion suns.

There is much more dark matter in the universe (which we cannot see) than there is luminous matter (which we can see). As explained in Appendix C, studies of the gravitational effects associated with galaxy motions show that there must be more than 300 times as much dark matter as luminous matter in the universe. Multiplying 60 billion, billion times 300 gives 18 billion, trillion. Thus the total estimated matter within our observable universe is equal to the mass of 18 billion times one trillion suns. (Eq. C-5)

Let us assume hypothetically that all of the matter within our observable universe (18 billion, trillion suns) is squeezed to form a single body with the density of water. The resultant body would have a diameter of 4.4 light years, which is approximately the distance to the nearest star. (Eq. C-12)

Now let us assume that this matter is squeezed further to achieve a single body with the density of nuclear matter, in accordance with the Gamow postulate. The resultant body would have a diameter of 700 million kilometers. This is about 1.5 times the diameter of the orbit of Mars (460 million kilometers). (Eq. C-13)

Thus, when we apply modern astronomical data to the Gamow Big Bang postulate, we find that the observable universe would have been about the size of the orbit of Mars at the instant of the Big Bang.

When Gamow presented his Big Bang theory, astronomers did not understand the role of supernovas in generating the heavy elements from which our solid earth is formed. Gamow postulated that the Big Bang explosion created these heavy elements. However, subsequent studies have convinced astronomers that supernovas are the primary source of heavy elements.

The Modern Big Bang Theory

As will be shown in Chapter 6, the very complicated equations of the Einstein general theory of relativity could be solved only for very simple cases during Einstein's lifetime. (Einstein died in 1955.) In the mid 1960's, powerful computers became widely available, and so a large number of scientists began to apply them to study the Einstein equations. With computers, the formidable Einstein equations could be solved in a manner unheard of in Einstein's day.

Most of these computer studies of the Einstein equations have been applied to cosmology, because that was about the only area that can use this expertise. About all of the cosmology studies predicted a Big Bang explosion at the beginning of the universe, and they generated an enormous amount of Big Bang research. This Big Bang computer research invariably predicted a *singularity* at the instant of the Big Bang, which ideally means that the density of matter was infinite and the size of the universe was zero.

This requirement of a singularity at the beginning of the universe expansion was expressed as follows in a book by Filkin [17] (p. 104), titled *Stephen Hawking's Universe*:

> *"Stephen [Hawking] and Roger Penrose published a paper in 1970 which proved that, if Einstein's mathematics were correct, a singularity had to result from a black hole, and had to exist at the start of the universe. - - - The paper argued that if relativity as explained by Einstein is correct — and all of the evidence from observation seems to keep confirming it — then the universe must have started with a big bang explosion out of a singularity. The equations do not allow an alternative."*

Filkin directed a 1997 television series for the Public Broadcasting System, also called *Stephen Hawking's Universe*. This book was a supplement to that television series.

How small was the "singularity" universe at the instant of the Big

5. Hubble Expansion of the Universe   97

Bang as predicted by the modern Big Bang theory?

James Peebles was called the "father of modern cosmology" by the January 2001 *Scientific American* (p. 37). The October 1994 *Scientific American* (page 53) presented an article by Peebles and others on the Big Bang theory. The article began with:

> *"At a particular instant, roughly 15 billion years ago, all of the matter and energy we can observe, **concentrated in a region smaller than a dime**, began to expand and cool at an incredibly rapid rate."*

Nearly all modern cosmologists endorse this claim by Peebles that the observable universe at the instant of the Big Bang was "smaller than a dime", and most believe it was very much smaller.

Most of the recent Big Bang theories support the *"inflation"* concept proposed by Alan Guth in 1981, which postulates that the initial universe was microscopic in size. Dickinson [28] (page 118) gives a pictorial representation of the Big Bang singularity that includes Guth's inflation postulate. **Dickinson states that during the inflation period, the observable universe "expands from one trillionth the size of a proton to the size of a baseball".** A proton is 2 trillionths of a millimeter in diameter.

Dickinson's book is a third edition of a popular astronomy book, with excellent illustrations and explanations. It is clear that the statements in this book portray the general thinking of astronomers.

Joseph Silk is the author of three books on the Big Bang theory. Silk [21] (p. 79) discusses two versions of the Big Bang theory. In an early Big Bang model, the observable universe began as a body that was one millimeter in diameter. However, Silk explains that most of the recent Big Bang models have endorsed the "inflation" concept of Alan Guth, in which the universe began with a diameter of 3 trillionths of a trillionth of a millimeter (3×10^{-24} mm).

The Modern Big Bang theory, with its singularity principle, is radically different from the Big Bang theory of George Gamow. Gamow assumed that the universe began with the density of nuclear matter (one billion tons per teaspoon), because he considered that this represented the greatest possible density of matter. The Gamow Big Bang theory predicts that the observable universe was initially about the size of the orbit of Mars around the sun. In contrast the modern Big Bang theory predicts that the observable universe was initially "smaller than a dime", and probably was microscopic in size.

These extreme predictions of the Modern Big Bang theory are

derived from studies of the Einstein general theory of relativity. Chapter 6 explains the Einstein theory, and Chapter 7 shows that the Einstein theory implies a singularity at the start of the universe expansion. Einstein recognized this point, but absolutely rejected all singularity predictions derived from his theory.

Cosmic Microwave Background Radiation

An important milestone in the development of the Big Bang theory occurred in 1965, when Arno Penzias and Robert Wilson, two physicists working at Bell Laboratories, discovered cosmic microwave background radiation. These physicists were performing measurements on a sensitive microwave antenna that had been developed for use in satellite communication. When communication satellites generated greater power, there was no longer a communication need for this sensitive instrument, and so the antenna was redirected toward basic research.

Penzias and Wilson discovered spurious signals in their antenna at microwave frequencies, which they could not explain. These signals were due to cosmic radiation coming from all directions in space. Big Bang theorists had predicted that such radiation should have been generated by the Big Bang.

Gamow had predicted that optical radiation from the early universe should still be observable today. With the expansion of the universe, this cosmic optical radiation should be reduced in frequency, and should be observable today at microwave frequencies. He predicted cosmic radiation coming from all directions that is equivalent to the radiation from an ideal blackbody. Gamow made several estimates of the effective temperature of this cosmic blackbody radiation, which varied from 5 degrees Kelvin to 20 degrees Kelvin.

The radiation from an ideal blackbody was discussed in Chapter 4. A general plot for the spectrum of blackbody radiation was given in Fig. 4-2 on page 74. An equation associated with Fig. 4-2 shows that the intensity of radiation varies as the fourth power of the absolute temperature. Consequently, the 4-to-one temperature range from 5 to 20 degrees Kelvin predicted by Gamow corresponds to an intensity ratio of 4 raised to the fourth power (4x4x4x4), which is 256. Thus the Gamow prediction of blackbody radiation was not very precise. There was an uncertainty factor of 256 in the predicted intensity of the blackbody radiation. (Intensity is received power per unit area.)

In the early 1960's, Prof. Robert Dicke of Princeton University was studying the Big Bang theory. One of his graduate students, James

Peebles, investigated Gamow's prediction of cosmic microwave radiation, and considered building an antenna to measure this radiation. Peebles estimated that the blackbody temperature should be 30 degrees Kelvin. Then he discovered that Penzias and Wilson had already measured similar radiation in their antenna. The antenna had detected unexplained electrical disturbance signals with a spectrum that corresponded to a blackbody temperature of about 3.5 degrees Kelvin.

This discovery of cosmic microwave background radiation was loudly publicized by Big Bang theorists. Since this cosmic radiation had been predicted by the Big Bang theory, it was proclaimed to be proof that the Big Bang theory was correct. However, the Big Bang proponents have failed to mention that this radiation was only predicted in a qualitative sense. Since radiation intensity varies as the fourth power of temperature, the 30 degree Kelvin estimate by Peebles corresponds to a radiation intensity that is more than 5000 times greater than the intensity for a blackbody at 3.5 degrees Kelvin.

In 1948, Ralph Alpher and Robert Herman, two graduate students working with George Gamow, had published a paper in the scientific journal *Nature* that predicted cosmic microwave radiation corresponding to a blackbody temperature of 5 degrees Kelvin. Since this is the closest estimate to the measured temperature, it is the value generally quoted by Big Bang proponents.

Much more accurate measurements of this cosmic radiation were obtained from the Cosmic Background Explorer (COBE) satellite in 1989. This satellite measured cosmic microwave radiation coming uniformly from all directions that corresponded very accurately, in intensity as well as in spectrum, to the radiation from an ideal blackbody at a temperature of 2.73 degrees Kelvin.

For a blackbody temperature of 2.73 degrees Kelvin, an equation on page 75 shows that the half-power wavelength of the blackbody spectrum is 1.50 millimeters, which is about 2500 times the peak wavelength of visible light. It corresponds to a frequency of 67,000 megahertz, which is about 100 times the frequency used for television.

The Big Bang theorists claim that this cosmic microwave background radiation is the cooled relic of optical radiation that was emitted from the early universe about 300,000 years after the Big Bang. However, an important weakness of this assumption is that the cosmic radiation emanates with extreme uniformity from all directions. Lerner [16] (p. 31) reports that this cosmic radiation is far too uniform to explain the lumpy arrangement of galaxies in the universe that has been observed in recent universe surveys. In the data obtained from the COBE satellite, the energy of radiation received from different directions varies by only a few parts in 100,000. The extreme uniformity of the COBE data with direction is examined in detail by Hoyle, Burbidge, and Narlikar [19] (pp. 89-94).

This implies extreme uniformity of the universe at this early period. How did this highly uniform early universe create what we observe today? Not only is our universe separated into galaxies, but the galaxies are not spaced uniformly. As we saw earlier on page 93, the galaxies are arranged into long curling filaments, which form huge ribbon-like structures that are typically 100 million light years thick and 300 million light years wide.

At the time that cosmic microwave background radiation was discovered by Penzias and Wilson in 1965, both the Steady State theory and the Big Bang theory were seriously considered by scientists as possible explanations for the Hubble expansion. The Steady State theory had been proposed in 1948 by Fred Hoyle, Hermann Bondi, and Thomas Gold. This theory did not have an immediate explanation for this cosmic radiation, and so interest in the Steady State theory declined rapidly.

Appendix D shows that a Steady State cosmology theory based on the Yilmaz theory of gravity also predicts cosmic microwave background radiation. The predicted blackbody temperature lies in the range between 2.1 and 3.4 degrees Kelvin. This range is much closer to the 2.73 degree Kelvin temperature measured by the COBE satellite than the best estimate (5 degrees Kelvin) made by Big Bang theorists.

Chapter 6

The Einstein Theory of Relativity

Basis for the Singularity Concept

The Gamow Big Bang theory gives a reasonable model of the universe that is consistent with our knowledge of the properties of matter. It assumes that at the beginning of the universe expansion the universe had the density of nuclear matter, which is the density of a neutron star (one billion tons per teaspoon). The eminent nuclear physicist, George Gamow, considered this to represent the greatest possible density of matter.

Why have modern Big Bang cosmologists rejected the Gamow theory to invoke their "singularity" concept, which predicts a nearly infinite density of matter? Why do they insist that the universe was squeezed from the size of the orbit of Mars to "the size of a dime", or probably to "one trillionth of the size of a proton"? The reason is that modern Big Bang theories are based on computer studies of the equations of Einstein's general theory of relativity. These equations predict that a singularity existed at the start of the universe expansion.

This chapter describes the Einstein theory. Despite the great complexity of the Einstein equations, the following shows that the principles of the Einstein theory can be described in a simple manner.

The basic Einstein theory of relativity (later called the *Special theory of Relativity*) was presented in 1905 to explain an enigma in measuring the speed of light. Hence our presentation begins by describing the principles of light propagation.

The Nature of Light

What Is a Light Wave?

We know that light and sound are waves. We can help to understand these waves by examining the propagation of a wave on water.

Mechanical Waves. We can envision a wave by dropping a stone into a smooth pond. The water waves propagate outward from the point where the stone hits the water. An individual water particle oscillates back-and-forth and up-and-down, following an elliptical path. It is the energy of the wave that propagates across the pond, not the water itself. This effect can be observed by noticing that an object floating on the surface moves very little as the wave passes it.

If we drop two stones into the pond at once, two sets of waves are produced, and these two sets of waves interfere with one another. Similar wave interference occurs with sound and light waves.

Like a wave on water, sound is a mechanical vibration. One can feel the vibration in a musical instrument when a tone is produced. The instrument vibrates the air, producing sound that travels through the air to the ear that hears it. Sound is a compression wave, in which the air molecules vibrate back and forth in the direction of propagation of the sound.

Electromagnetic Waves. A wave on water and a sound wave are mechanical waves, which propagate by vibrating a medium. Although a light wave is similar in certain respects to these mechanical waves, it is fundamentally very different. A light wave is a packet of oscillating electric and magnetic fields, and so is called an electromagnetic wave.

A light wave is the same as a radio wave except that it oscillates at a much higher frequency. Standard AM (amplitude modulation) radio operates at a frequency of about one megahertz (MHz). One hertz (Hz) means one cycle per second, and so one megahertz means one million cycles per second. Television operates at a frequency of about 500 megahertz; light has a frequency of about 500 million megahertz, and an X-ray has a frequency of about 3 trillion megahertz.

To help understand an electromagnetic wave, let us consider some examples of electric and magnetic fields. We observe the effect of an electric field in a thunderstorm. The severe winds of the storm remove electrons from the ground and deposit them in the clouds. The clouds are charged negatively with respect to the ground, and a large electric field builds up between the clouds and the ground. Eventually this electric field gets so strong that an electric spark (called lightning) jumps between a cloud and the ground. Electrons flow from the cloud to the ground, and temporarily eliminate the electric field.

Our earth has a magnetic field, which we can sense with a magnetic

6. The Einstein Theory of Relativity 103

compass. A magnetic compass is a magnet that is free to turn, and points in the direction of the earth's magnetic field. Our earth acts like a huge magnet, which has north and south poles that are close to the poles about which the earth rotates. Consequently a magnetic compass points approximately in the direction of the North Pole.

Although electric and magnetic fields are quite different, they are closely related, as shown by the following:

(1) A changing magnetic field produces an electric field
(2) A changing electric field produces a magnetic field

Property (1) can be observed in an automobile generator, which delivers electric current to charge the battery. The rotating part of the generator, called the rotor, acts as a magnet. As the rotor spins, its magnetic field moves through a coil of wire on the fixed part of the generator (the stator). This action produces an electric field that generates an electric current to charge the battery. *Hence, by property (1), a changing magnetic field produces an electric field.*

Property (2) is observed in an electromagnet. Flowing electrons produce an electric current, which is a moving electric field. Feeding the current through a coil of wire forms an electromagnet, which generates a magnetic field. The current in the coil is a moving (or changing) electric field, and this changing electric field produces a magnetic field. *Hence, by property (2), a changing electric field produces a magnetic field.*

Principle of an Electromagnetic Wave. An electromagnetic wave consists of oscillating (or changing) electric and magnetic fields. By property (1) the oscillating magnetic field produces the oscillating electric field, and by property (2) the oscillating electric field produces the oscillating magnetic field. Hence the oscillating electric and magnetic fields support one another, to form a packet of energy that propagates as a wave at the speed of light.

The oscillating electric and magnetic fields are perpendicular to one another. They produce an electromagnetic wave that propagates in the direction perpendicular to the plane of the electric and magnetic fields.

For example, assume that the electric and magnetic fields lie in the horizontal plane, with the magnetic field oscillating in the north-south direction and the electric field oscillating in the east-west direction. This means that half of the time the magnetic field points north, and half of the time it points south. Similarly, the electric field points east half of the time and points west half of the time.

The resultant electromagnetic wave propagates in the vertical direction. Whether the wave propagates upward or downward depends on the relative timing of the electric and magnetic fields. The electric and magnetic fields are out of phase with one another. This means that when the magnetic field is maximum, the electric field is zero; and when the electric field is maximum, the magnetic field is zero.

Discovery of the Radio Wave

In the 1700's and 1800's, extensive experiments on electricity and magnetism were performed by many scientists. This included Benjamin Franklin's experiments with lightning, which proved that lightning is electricity. Batteries were constructed that generated electricity. Electricity was applied to solutions in order to electroplate metals and to separate water into oxygen and hydrogen. Electric current was fed through a coil of wire to produce a magnetic field. Rotating magnetic fields produced electric motors and electric generators.

In 1873, James Clerke Maxwell (1831-1879) derived a set of mathematical equations that combined electric and magnetic effects into a mathematically consistent theory. These are called Maxwell's electromagnetic field equations. These equations are used as the theoretical basis for the design of countless devices, such as radio and radar antennas, microwave wave-guides, and micro-electronic circuits.

Maxwell predicted from his electromagnetic field equations that an electromagnetic "radio" wave could be generated. From measured values of electric and magnetic parameters, Maxwell calculated the speed of propagation of his electromagnetic wave, and found that it was equal to the measured speed of light. Therefore, Maxwell concluded that light must be an electromagnetic wave of very high frequency.

In 1888 (15 years after Maxwell presented his equations, and 9 years after Maxwell's death) Heinrich Hertz (1857-1894) performed the first radio propagation experiment. Hertz was able to transmit a radio wave across his laboratory.

Seven years later, in 1895, Guglielmo Marconi (1874-1937) used a radio wave of much lower frequency in the first practical radio communication system, when he was only 21. (*Guglielmo* is Italian for *William* and is pronounced *"Gulliamo"*.) Marconi performed his first radio experiment when he was 16, only 2 years after the experiment by Hertz.

Marconi formed a company to develop his radio system. He transmitted radio across the English Channel in 1898, from England to

Newfoundland in 1901, and from England to Cape Cod, Massachusetts in 1903. His antenna on Cape Cod was supported on four enormous towers, 200 feet tall. The antenna consisted of wires that formed a conical mesh that rose from the building below to the top of the towers.

The Aether Concept

The ancient Greeks knew that sound propagates as a wave, but what is light? Is light a wave like sound, or does it consist of particles, shot out like stones from a slingshot?

Sound propagates by vibrating a mechanical medium, such as the wood of a violin or the molecules in the air. Many people believed that light propagates by vibrating a mysterious substance called the aether, which fills all of space. Isaac Newton rejected the wave concept of light because the aether did not make sense. He asked, "How can the aether be so thin that it has no effect on the motion of planets, yet be so stiff that it allows light to propagate at a tremendous speed?" Based on this reasoning, Newton concluded that light was a stream of particles. [30]

In the early 1800's, Thomas Young performed light interference experiments, which proved that light is a wave. He was publicly ostracized for having the audacity to doubt the great Sir Isaac Newton. [71] However, in 1819, Augustin Fresnel performed more accurate light interference experiments that were accepted as proof that light is a wave.

But how does light propagate? The mysterious aether that allows light propagation was as enigmatic as ever.

When Maxwell presented his equations in 1878, the mystery of light propagation was explained. Light does not propagate by vibrating an aether medium. Light consists of oscillating electric and magnetic fields, which form packets of energy that propagate at the speed of light.

Even though the aether made no sense physically, and was not needed to allow light propagation, Maxwell still included the aether in his theory. Maxwell assumed that the aether acts as a reference medium relative to which the light propagates. The need for an aether becomes clear when one considers the problem of measuring the speed of light.

Measuring the Speed of Light

The Speed of Sound. To introduce the issue of measuring the speed of light, let us consider the much simpler problem of measuring the speed of sound. Sound travels relative to the air at a speed of 330 meter/sec (meters per second). Let us assume that the wind is blowing at 20 meter/sec. An instrument is placed on the ground to measure the speed of the sound wave. If the wind is blowing against the sound wave, the speed of the sound measured by the instrument on the ground would be (330 - 20) or 310 meter/sec. If the wind is blowing in the direction of the sound wave, the speed of the sound measured by the instrument on the ground would be (330 + 20) or 350 meter/sec.

The Speed of Light. Light travels about one million times faster than sound, at a speed of 300 million meters per second, or 300,000 kilometers per second (km/sec).

What do we mean by the speed of light? Do we measure the light speed relative to the emitter that generates the light, or relative to the receiver that receives the light? If there is a large velocity between the emitter and the receiver, the values of the speed of light can differ greatly for the two assumptions.

For this reason, many scientists believed that there must be an aether medium relative to which the light propagates. Light would be measured relative to the aether, just as sound is measured relative to the air. However, many attempts were made to measure the speed of the aether "wind", and all of them failed completely.

Let us see what happens when one measures the speed of light emitted by a star that is moving at an appreciable velocity relative to the earth. As shown in Fig. 6-1, a light wave is radiated by a star (the light emitter) and is sensed by a light receiver on the earth. Assume that the star is moving at a velocity of 300 km/sec toward the earth, and that the earth is moving at a velocity of 200 km/sec toward the star, so that the relative velocity between the star and the earth is 500 km/sec.

When the speed of a light wave is measured relative to a light emitter, it is always 300,000 km/sec. Hence one would expect that the light emitted by a star should travel at 300,000 km/sec relative to the star. Since the star is traveling at 300 km/sec, one would expect the absolute velocity of the light to be the velocity sum (300,000 + 300), which is 300,300 km/sec. These velocities are all shown in Fig. 6-1.

Since the earth is moving toward the light wave at 200 km/sec, the velocity of the light relative to the earth should presumably be the sum (300,300 + 200) or 300,500 km/sec. Thus one would expect the velocity of the light relative to the earth (the receiver) to be equal to the nominal speed of light (300,000 km/sec) plus the relative velocity between the star and the earth (500 km/sec for this example).

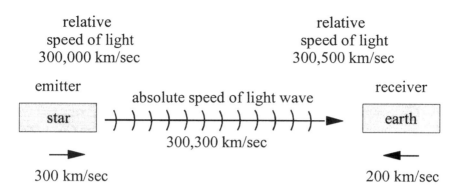

Figure 6-1: Assumed speed relations associated with the propagation of a light wave from a star to earth

But that is not what happens! The speed of the light received from any star is always exactly the same, 300,000 km/sec. Nevertheless, the spectral shift of the light from a star shows that the star's velocity can vary greatly from star to star. The speed of light measured relative to a light receiver, or relative to a light emitter, is always exactly 300,000 km/sec. The speed of light is completely independent of the velocity of the emitter or the velocity of the receiver.

How can we explain this constancy of the speed of light? During the late 1800's, scientists were greatly perplexed by this enigma.

The Einstein Relativity Principle

Into this confusion stepped the brilliant young physicist, Albert Einstein (1879-1955). In 1905 Einstein [8] published his paper on Relativity. Einstein approached the enigma of the speed of light by applying fundamental reasoning. He concluded that *absolute velocity* has no meaning; there is only *relative velocity*. **We can specify the relative velocity between two bodies, but not the absolute velocity of either one.**

Based on this principle, Einstein concluded that the aether cannot exist. Since absolute velocity has no meaning, there cannot be an aether medium that establishes an absolute reference for specifying velocity.

How do we specify the speed of light? In Fig. 6-1, the light travels from the emitter (star) to the receiver (earth). Since the speed of light can only be specified in a relative sense, we can consider (1) that the speed of light is the relative velocity between the emitter and the light, and (2) that the speed of light is the relative velocity between the receiver and the light.

Since absolute velocity has no meaning, the velocity of an observer located at the emitter is no more (and no less) preferred than the velocity of an observer located at the receiver. Consequently, observers located at the emitter and receiver must measure exactly the same value for the speed of light.

Thus, Einstein established the following principle: ***Two observers moving at different velocities must measure exactly the same value for the speed of light, regardless of the relative velocity between them.***

Einstein then asked, "What conditions must be satisfied in order for this principle to hold?" Einstein concluded that if there is a relative velocity between two observers, their measurements of distance and time must be different. [8]

This means, for example, that an object does not have a definite physical length. The length of an object depends on the velocity of the observer that is measuring it. Dimensions are relative. Time intervals are relative. **Reality is Relative**. This is the essence of the Einstein theory of Relativity.

In 1905, when Albert Einstein published his paper on relativity, he was an unknown physicist working in the Swiss Patent Office. In that same year Einstein published three other landmark papers. [23] (pp. 120-121) One paper dealt with the photoelectric effect, which proved that light is quantized into small units, called photons. Another paper determined the true size of molecules from their diffusion in a diluted liquid solution. A third paper analyzed the statistics of the motions of gas particles, which established Einstein as the founder of statistical mechanics. Einstein's relativity paper was the fourth of these landmark papers on physics published in 1905.

In 1922, Einstein received the Nobel Prize for his paper on the photoelectric effect. He did not receive the Nobel Prize for his much more important research on relativity, because it was too controversial at the time. Einstein's paper on the photoelectric effect proved that light not only acts like a wave; it also acts like a stream of particles.

The Einstein Special Theory of Relativity

Equipment for Measuring the Speed of Light

In presenting his theory, Einstein first considered the steps involved in measuring the speed of light. This is achieved by measuring the time for light to travel the length of a measuring rod. Let us assume that the measuring rod is 3 meters long.

Light travels 300 million meters per second. This can be expressed as 300 meters per microsecond, where a microsecond is one millionth of a second. In 1/100 microsecond, light travels 3 meters. Hence light travels the length of a 3-meter measuring rod in 1/100 microsecond.

It is convenient to express this time in terms of the nanosecond, which is one thousandth of a microsecond, or one billionth of a second. A time interval of 1/100 microsecond is equal to 10 nanoseconds. Hence light travels the length of a 3-meter measuring rod in 10 nanoseconds.

Figure 6-2 shows the process for measuring the time for light to travel the length of a 3-meter measuring rod. Clocks 1 and 2 are placed at the two ends of the measuring rod. The two clocks run at exactly the same rate, and are accurately synchronized. Synchronization could be achieved by locating the clocks at the same point, setting their readings equal, and then moving them to the ends of the measuring rod.

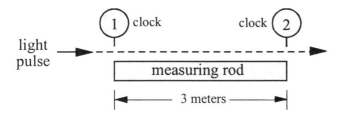

Figure 6-2: Measuring the speed of light

When a light pulse reaches the leading edge of the measuring rod, clock 1 is read. When the light pulse reaches the trailing edge, clock 2 is read. The two clock readings are subtracted to obtain the time for light to travel the length of the 3-meter measuring rod. The difference in the readings of clocks 1 and 2 is exactly 10 nanoseconds.

Einstein concluded that when observers moving at different velocities perform this experiment of measuring the speed of light, their measurements appear to be inconsistent, because (1) their measuring rods appear to have different lengths, (2) their clocks appear to run at different rates, and (3) clocks that appear to be synchronized relative to one observer are not synchronized relative to the other observer.

Application of Relativity to a Fictitious Space-Travel Experiment

Let us apply the Einstein relativity theory to a fictitious space travel experiment, which involves sufficient velocity to produce strong relativistic effects. As shown in Fig. 6-3, a space ship is traveling toward earth at 60 percent of the speed of light. Since the speed of light c is 300 meters per microsecond, the space ship velocity V is 180 meters per microsecond.

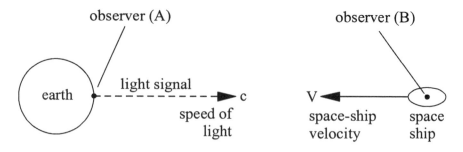

Figure 6-3: Measuring the speed of a light signal transmitted from earth to a space ship, which is returning at a velocity V equal to 60 percent of the speed of light c.

A light pulse is transmitted from earth to the space ship. Relative to the earth observer A, the light pulse travels at the normal speed of light c, which is 300 meters per microsecond. To the earth observer A, this light pulse travels relative to the space ship at the velocity (300 + 180) or 480 meters per microsecond. (This is the sum of the normal speed of light c plus the velocity V of the space ship.)

Einstein postulated that, to the earth observer A, the measuring rod on the space ship appears to be compressed by the following factor K

$$K = \sqrt{[1 - (V/c)^2]} = \sqrt{[1 - (0.6)^2]} = \sqrt{[0.64]} = 0.8$$

The parameter V is the space ship velocity and c is the speed of light. Since the space ship is traveling at 60 percent of the speed of light, the ratio V/c is equal to 0.6. The calculation shows that the compression factor K for this experiment is equal to 0.8.

To the earth observer, the measuring rod on the space ship appears to be compressed to 80 percent of its normal length. A clock rate is reduced by this same compression factor K. To the earth observer, the clocks on the space ship appear to run at 80 percent of their normal rates.

The space ship observer B considers his measuring rod to be 3 meters long. To the earth observer A, this measuring rod is compressed to 80 percent of its normal (3 meter) length, to a length of 2.4 meters.

6. The Einstein Theory of Relativity   111

The space ship observer B considers his two clocks to be accurately synchronized, but the earth observer A considers the two clocks to be out of synchronization by the following amount:

Synchronization error = $(V/c)\Delta t$ = 0.6 Δt

The quantity Δt is the time for light to travel between the two clocks, as seen by the space ship observer B. Observer B finds that it takes 10 nanoseconds for light to travel the length of his measuring rod, and so Δt is 10 nanoseconds. Observer A sees a clock synchronization error that is 60 percent of 10 nanoseconds, which is 6 nanoseconds. This synchronization error is measured in terms of the space-ship clocks.

Observer B on the space ship considers his two clocks to be exactly synchronized, but observer A on earth considers these two clocks to be out of synchronization by 6 nanoseconds, as measured on the space-ship clocks.

To observer A on earth, the light pulse appears to travel relative to the space ship at a velocity of 480 meters per microsecond, which is the normal speed of light c plus the space ship velocity V. The measuring rod on the space ship appears to have a length of 2.4 meters (80 percent of the normal length). Hence, relative to the earth observer A, the time for light to travel the length of the space-ship measuring rod is 2.4/480 or 1/200 microsecond, which is *5 nanoseconds*. This time interval is the apparent rod length (2.4 meters) divided by the apparent speed of the light pulse relative to the rod (480 meters per microsecond).

Thus, to the earth observer A, the light takes 5 nanoseconds to travel the length of the measuring rod. However, the clocks on the space ship appear to run at 80 percent of the normal rate. Hence the space ship observer B should observe a time interval that is 80 percent of 5 nanoseconds, which is 4 nanoseconds. *Relative to the earth observer A, it will take 4 nanoseconds for the light pulse to travel the length of the space ship measuring rod, as measured on the space-ship clocks.*

To the earth observer A, the clocks on the space ship are out of synchronization by 6 nanoseconds. This 6-nanosecond synchronization error should be added to the 4-nanosecond time interval for the light pulse to travel the length of the measuring rod. This sum gives the measured time interval for light to travel the length of the measuring rod, which is 10 nanoseconds. Hence the earth observer A recognizes that the space ship observer B should find that it takes 10 nanoseconds for light to travel the length of the space-ship measuring rod.

In this manner, we can explain why both observers A and B find that it takes exactly 10 nanoseconds for the light pulse to travel the length of a measuring rod in their equipment. Consequently both observers measure exactly the same value for the speed of light.

Symmetry of Relation between Observers A and B.

Einstein concluded that absolute velocity does not exist; there is only relative velocity. The earth observer A can assume that the earth is stationary and the space ship is moving at 60 percent of the speed of light. Similarly, the space ship observer B can assume that the space ship is stationary, and the earth is moving at 60 percent of the speed of light.

Let us assume that the space ship sends a light pulse to earth. The preceding discussion can then be applied to the space ship rather than to the earth, by considering the space ship observer to be A and the earth observer to be B. The measurements of the speed of light are symmetric.

Implications of Relativistic Effects

We have seen that, relative to the earth observer, the space ship measuring rod appears to have 80 percent of its normal length, the space ship clocks appear to run at 80 percent of the normal rate, and the two space ship clocks appear to be out of synchronization by 6 nanoseconds, even though the space ship observer considers his clocks to be accurately synchronized.

According to the principle of relativity, these apparent effects are real. Distance and time measurements are not absolute. They depend on the velocity of the observer. The principle that **Reality is Relative** has profound physical implications.

The Fourth Dimension. The finding that clocks synchronized on the space ship are not synchronized relative to the earth observer shows that time is not absolute. If two synchronized clocks are located at the same point, they are synchronized relative to all observers; but when they are located at different points, they are not necessarily synchronized to two observers moving at different velocity.

This indicates that measurements in time and space cannot be considered separately. In relativity, time and spatial measurements must be combined together into a *four-dimensional space-time specification*. This does not mean that we should regard time to be a mysterious fourth dimension that is equivalent to a spatial dimension. To any observer, space and time measurements are radically different concepts.

6. The Einstein Theory of Relativity 113

The fourth-dimensionality of space-time means that time and space must be considered together to obtain a precise specification. One cannot separate a measurement of time from a measurement of distance. For example, a time interval between two events that is experienced by one observer can appear to be a distance interval to another observer moving at a different velocity.

Conversion of Matter into Energy. From his relativity principle, Einstein derived some profound physical conclusions. He showed that mass M and energy E can be converted into one another according to the famous Einstein formula:

$$E = Mc^2$$

This formula explained the source of the energy radiated by the sun, and eventually led to the atomic and hydrogen bombs. The equation shows that one gram of mass M is equivalent to 25 million kilowatt-hours of energy E. This means that 25 million kilowatt-hours of energy are released for every gram of matter that is converted into energy. One gram has 1/3 of the weight of a United States penny (one-cent coin).

The sun generates its energy by fusing four hydrogen atoms to form one helium atom. The helium atom has 0.710 percent less mass than the four hydrogen atoms that form it. The mass that is lost is converted into energy. Taking 0.710 percent of 25 million kilowatt-hours gives 177,500 kilowatt-hours. Therefore, when the sun fuses one gram of hydrogen to form helium, it generates 177,500 kilowatt-hours of energy.

It is remarkable that the abstract principles of special relativity allowed Einstein to explain the source of the enormous energy radiated by our sun. ***The fact that matter can be converted into energy, in accordance with this Einstein formula, demonstrates that Einstein's Relativity principles are correct. It proves that Reality is Relative.***

Variation of Mass with Speed. Another important implication of relativity is that the mass of an object increases with its speed, as shown by the following formula:

$$Mass = (Rest\ Mass)/\sqrt{[1 - (V/c)^2]}$$

Rest mass is the mass of the object when its speed is zero. A plot of this equation is shown in Fig. 6-4. This plot shows that the mass of an object approaches infinity as its velocity approaches the speed of light. This has been verified by accelerating electrons in an electric field. *Figure 6-4 shows that nothing with mass can travel at the speed of light.*

For those who dream of hopping around our galaxy, as depicted in the *Star Trek* television series, this plot demonstrates a severe limitation. Regardless of how much our technology may advance, the velocity of a space ship can never reach the speed of light.

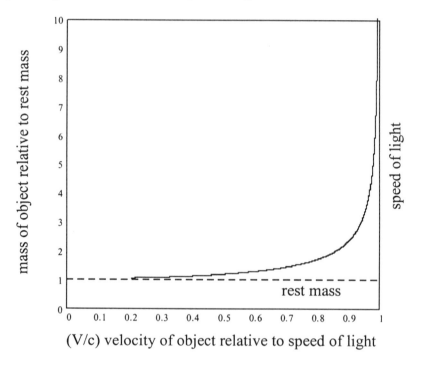

Figure 6-4: Increase of mass of an object with velocity; its mass becomes extremely high as its velocity approaches the speed of light

The Einstein General Theory of Relativity

Generalizing the Relativity Principle

After presenting his basic theory of relativity in 1905, Einstein discovered that it had a serious theoretical weakness. His theory applies exactly only when the velocity is constant. When acceleration occurs, which means that the velocity is changing, he discovered that his equations do not hold exactly. He realized that acceleration and gravity are equivalent, and so his basic relativity theory does not apply exactly in a gravitational field.

To achieve a rigorous (mathematically consistent) theory, it was essential that Einstein generalize his relativity theory to include the effects of acceleration and gravity. Einstein struggled for 11 years until he achieved his *General theory of Relativity* in 1916.

Einstein began the generalization of his theory by performing approximate calculations of the effects of acceleration and gravity on time and distance measurements.

Equivalence between Acceleration and Gravity

To calculate the relativistic effects of gravity, Einstein postulated that acceleration and gravity are equivalent. The gravitational pull of the earth exerts a force on our bodies, which we call weight. In an airplane during take-off, the body is forced back against the seat because of the forward acceleration of the airplane. Einstein postulated that the relativistic effects of gravity and acceleration forces are equivalent. To generalize his relativity theory, he applied the principles of relativity under conditions of acceleration, and related these results to a gravitational field. Let us see how he did this.

In Fig. 6-5, Einstein considered two identical elevators. One in diagram (b) rests on the ground, and the other in (a) is located in space, where gravitational force is negligible. Einstein assumed that the elevator in space is being pulled upward with the same acceleration of gravity g that is experienced on earth. Today we assume more realistically that a rocket motor under elevator (a) is pushing the elevator upward with the acceleration g.

The meaning of the acceleration of gravity g was explained in Chapter 3. When an object is allowed to fall on earth, it drops with a constant acceleration of gravity g of about 10 meter/sec per second. The velocity of a falling object is proportional to time, provided that air resistance is negligible. The velocity is 10 meter/sec in one second, 20 meter/sec in two seconds, etc. When an object is not allowed to fall, it exerts a force on the floor, which is called weight. The weight of an object is equal to its mass multiplied by the acceleration of gravity g.

Since the elevator in diagram (a) is pushed upward by a rocket with the same acceleration of gravity g experienced on earth, an object within elevator (a) would exert the same weight force on the floor that it would exert if it were located within the fixed elevator (b) located on earth. The conditions inside elevators (a) and (b) are the same. If the elevators are closed, a scientist could not tell from experiments performed within the two elevators whether he is in elevator (a) or elevator (b). This is the *Principle of Equivalence* proposed by Einstein, which allowed him to relate the effects of acceleration and gravity.

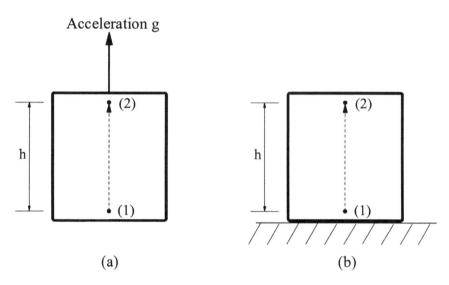

Figure 6-5: Elevator (a) is accelerating in free space; elevator (b) is fixed on earth. Light travels from point (1) to point (2).

Redshift Produced by Gravity

Suppose that a light pulse is emitted from the floor of elevator (a) at point (1) and is received at the ceiling at point (2). During the propagation time of the light, the velocity of the elevator increases, because the elevator is accelerating. Therefore the upward velocity of the receiver at point (2) is greater than that of the emitter at point (1), as far as the light pulse is concerned. Point (2) appears to be moving away from point (1), and so the light is redshifted as it moves from (1) to (2). The wavelength of the light received at (2) is greater than the wavelength emitted at (1).

The time for light to travel from the floor to the ceiling of the elevator is approximately equal to the height h of the ceiling above the floor divided by the speed of light c. This time interval is (h/c).

The time interval (h/c) is multiplied by the elevator acceleration (g) to obtain g(h/c), which is the velocity difference ΔV between points (1) and (2). Dividing this velocity difference ΔV by the speed of light c gives the velocity ratio $\Delta V/c$, which is

$$\Delta V/c = gh/c^2$$

The wavelength of the light is denoted L, the wavelength shift is ΔL, and the relative wavelength shift is the ratio $\Delta L/L$. This wavelength ratio is approximately equal to the velocity ratio $\Delta V/c$, and so is approximately equal to

$$\Delta L/L = \Delta V/c = gh/c^2$$

The wavelength ratio $\Delta L/L$ is called the redshift of the light. This analysis shows that when light travels from the floor to the ceiling of elevator (a) the light experiences a redshift of approximately gh/c^2.

Conditions in elevators (a) and (b) are identical. Consequently, the gravitational field in the fixed elevator (b) on the ground produces a $\Delta L/L$ redshift approximately equal to gh/c^2 as the light propagates from the floor to the ceiling of the elevator.

Effect of Gravity on a Time Measurement

Assume that a light signal is used as the timing reference for a clock, to produce a clock tick every cycle of the light wave. Clocks are located at the floor and ceiling of the elevator. To synchronize the two clocks, a light signal is transmitted from the floor to the ceiling of the elevator.

The clock period T (the time between clock ticks) is proportional to the wavelength of the light. Because of redshift as the light propagates from the floor to the ceiling, the clock at the ceiling has a longer clock period T than the clock on the floor. The difference of the two clock periods is denoted ΔT. The ratio $\Delta T/T$ is equal to the redshift $\Delta L/L$ of the light. Hence the relative difference of the two clock periods is approximately equal to

$$\Delta T/T = \Delta L/L = gh/c^2$$

A clock is the basis for performing a time measurement. This equation shows the approximate change of a *time measurement* that is produced by gravity or acceleration.

Effect of Gravity on a Distance Measurement

Einstein performed another analysis showing that gravity causes a spatial dimension to contract. The relative decrease of a spatial dimension (a *distance measurement*) is approximately equal to

$$\Delta x/x = gh/c^2$$

where x is a spatial dimension, and Δx is the decrease of the dimension.

Effect of Gravity on the Speed of Light

The changes of *time and distance measurements* showed that gravity causes the *speed of light* to decrease. The relative decrease in the speed of light is approximately equal to:

$$\Delta c/c = (\Delta T/T) + (\Delta x/x) = 2gh/c^2$$

Ratio ($\Delta c/c$) is the relative decrease of the speed of light, ($\Delta T/T$) is the relative increase of a clock period, and ($\Delta x/x$) is the relative decrease of a spatial dimension. The ($\Delta T/T$) increase of a clock period can also be expressed as an equal decrease in the relative clock rate.

Thus, Einstein found that gravity (and acceleration) produce a relative reduction of the speed of light approximately equal to $2(gh/c^2)$, and produce relative reductions of a clock rate and a spatial dimension approximately equal to (gh/c^2).

Constancy of the speed of light was the principle on which Einstein based his *special theory of relativity*. Since the speed of light is not always constant, Einstein needed a new principle to generalize his relativity theory so that it applies under conditions of gravity and acceleration.

The Mathematical Theory of Curved Space

Einstein needed to combine into a rigorous theory the relativistic effects produced by gravity and acceleration, with the relativistic effects produced by velocity (given in his special theory). Einstein achieved this generalization of relativity by applying the complex mathematics of curved space that had been developed by Riemann and Ricci, which is now called *tensor analysis*. Einstein characterized the effects of gravity and acceleration as a curvature of four-dimensional space. This allowed him to express his general theory of relativity in terms of the mathematics of curved space.

The foundation for tensor analysis was laid by the German mathematician Bernhard Riemann (1826-1866). In 1852 he presented the metric equation, which is a general mathematical principle for specifying curved space. The metric equation is based on the geodesic, which is the shortest distance between two points in curved space.

Riemann was unable to develop his concepts in detail, because he contacted tuberculosis in 1862, and died four years later at age 39.

The Italian mathematician, Gregorio Ricci (1853-1925), used the curved-space principle of Riemann as the foundation for a comprehensive mathematical theory, which Ricci called the *absolute differential calculus*, and is now called *tensor analysis*. *Tensor analysis* applies calculus to curved space.

Ricci [11] published his mathematical theory in 1901 with the help of his student, Tullio Levi-Civita (1873-1941). In 1923, Levi-Civita published in Italian an updated version of this theory. An English translation of this book is available as a Dover reprint. [7]

Unfortunately the monumental contributions of Ricci and Levi-Civita to general relativity theory have been largely ignored. The mathematical foundation for general relativity is commonly referred to as Riemannian geometry. However this foundation was really the calculus of curved space developed by Ricci, which was a major extension of the Riemannian geometric principle.

Einstein developed his general theory by incorporating into tensor analysis the principles of Newton's theory of gravity and the principles of relativity. This was a tremendous achievement. Einstein specified his theory in terms of a tensor formula called the **Einstein Gravitational Field Equation**. As explained in Appendix E, this tensor formula is very complicated, because it represents 10 independent equations, which are expressed in terms of the mathematical formulas specified by the complex Riemann-Ricci mathematics of curved space.

Verification of General Relativity

Because of the great complexity of the Einstein equations, Einstein was only able to derive approximate solutions from them. Karl Schwartzschild (1873-1916), who was cooperating with Einstein, applied the Einstein formula to a simple physical model of a star (or our sun), and thereby achieved an exact solution. Einstein also published the Schwartzschild solution in 1916. Unfortunately Karl Schwartzschild died suddenly from disease even before his famous analysis was printed. He was a German army officer at the Russian front during World War I.

The Schwartzschild solution was used as the basis for formulating tests to verify Einstein's general theory. General relativity incorporates the special relativity effects caused by velocity and the general relativity effects caused by gravity and acceleration. Within the weak gravitational fields of our solar system, the relativistic effects due to gravity are very small, but are measurable. From the Schwartzschild solution, Einstein devised the following three tests to verify the gravitational predictions of his general theory of relativity:

(1) When a light ray passes close to the sun, it should be deflected by 1.8 seconds of angle (arc seconds).
(2) A gravitational field causes a clock to run slower and a wavelength to increase. The gravitational field of our sun should cause the spectrum of light from the sun surface to shift toward the red by 2.1 parts per million of wavelength.
(3) The planet Mercury has a highly elliptical orbit. The axis of the Mercury orbit advances (or rotates) by 1.39 arc seconds per orbit. Of this advance of the orbit axis, 1.29 arc seconds can be explained with Newton's laws by considering the gravitational attraction of other planets. A residual error of 0.10 arc second per orbit remained, which was explained by the Einstein general theory of relativity.

These three tests were implemented, and the results established the validity of the Einstein general theory of relativity.

These measurable gravitational effects of general relativity are very small: an advance of only 0.10 arc second per orbit of Mercury; a 1.8 arc second deflection of a light beam passing close to the sun, and a gravitational redshift of only 2.1 parts per million in light emitted from the sun. Hence one might wonder why Einstein worked so hard to achieve his theory, and why general relativity is so highly regarded. The answer: *This generalization was essential to provide a solid theoretical foundation for the relativity principle embodied in special relativity.*

When the predictions of general relativity were verified, Einstein achieved great fame. After that time, Einstein did little with his general theory. His special relativity theory is very much easier to apply, and has wide applicability. During Einstein's lifetime, general relativity served primarily as a theoretical foundation for justifying special relativity.

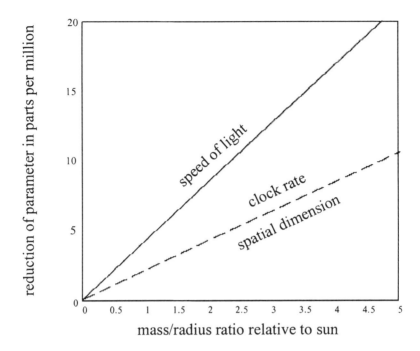

Figure 6-6: Relative reduction of speed of light, clock rate, and spatial dimension predicted by the Einstein theory, produced by gravity, expressed in terms of mass/radius ratio relative to the ratio at the surface of our sun

Reduction in Speed of Light, Clock Rate, and Spatial Dimension Produced by Gravity

Figure 6-6 shows the reductions of the speed of light, a clock rate, and a spatial dimension produced by gravity, which were predicted by the Schwartzschild solution to the Einstein theory. The gravitational field outside a spherically symmetric body is proportional to the mass/radius ratio at that point. This ratio is the mass of the body divided by the radial distance (radius) measured from the center of the body.

The reductions of the parameters in Fig. 6-6 are expressed in parts per million. One percent, which means one part per hundred, is equal to 10 parts per thousand, or 10,000 parts per million. Hence, 10 parts per million means 1/1000 of one percent (0.001 percent).

The horizontal scale in Fig. 6-6 expresses the mass/radius ratio relative to the value at the surface of the sun. A value of unity on the horizontal scale represents the surface of a star having the same mass/radius ratio as our sun.

For tests performed within our solar system, the relative mass/radius ratio is unity (at the surface of the sun) or less (outside the sun). Figure 6-6 shows that the relativistic effects due to gravity are extremely small for experiments performed within our solar system. At the surface of our sun (a value of 1.0 on the horizontal scale), Fig. 6-6 shows that the speed of light is reduced by 4.2 parts per million, and a clock rate (or a spatial dimension) is reduced by 2.1 parts per million.

The radius of the earth orbit (150 million kilometers) is 215 times greater than the radius of the sun (700 thousand kilometers). Hence the gravitational field of the sun at the earth's orbit is the value at the surface of the sun divided by 215. Thus the relative mass/radius ratio for the sun's gravity at the orbit of the earth is 1/215, which is approximately 0.005. Applying this value to the horizontal scale of Fig. 6-6 shows that, at the orbit of the earth, the sun's gravity reduces the speed of light by the infinitesimal amount of only 0.02 parts per million.

The Schwartzschild Limit

The percent reduction of the speed of light is proportional to the mass/radius ratio of the star. If a star has 240,000 times the mass/radius ratio of our sun, the speed of light would be reduced by 100 percent, and so the speed of light would be zero at the surface of the star.

If the mass/radius ratio of a star is greater than 240,000 times the ratio of our sun, the Schwartzschild solution does not yield an answer. (The Schwartzschild analysis predicts that the pressure inside the star would have an "imaginary" value, which is physically impossible.) For this reason, a mass/radius ratio of 240,000 times that of our sun is called the **Schwartzschild limit**. The Schwartzschild analysis does not yield an answer if the mass/radius value of a star exceeds the Schwartzschild limit.

Figure 6-7 presents the quantities shown in Fig. 6-6 for very large values of the mass/radius ratio. This ratio is expressed in terms of the Schwartzschild limit. A value of unity on the horizontal scale means that the mass/radius ratio of the star is equal to the Schwartzschild limit, which is 240,000 times the mass/radius ratio for our sun. These plots are derived in *Story* [2], Chapter 10 and Appendix E.

Figure 6-7 shows that all of the parameters (speed of light, clock rate, and spatial dimension) go to zero when the mass/radius ratio of a star reaches the Schwartzschild limit. (For mass/radius ratios exceeding the Schwartzschild limit, the Schwartzschild analysis does not yield an answer.) When the Einstein theory was validated (about 1920), the Schwartzschild limit far exceeded any known mass/radius ratio of a star. Consequently Einstein considered the Schwartzschild limit to be physically irrelevant.

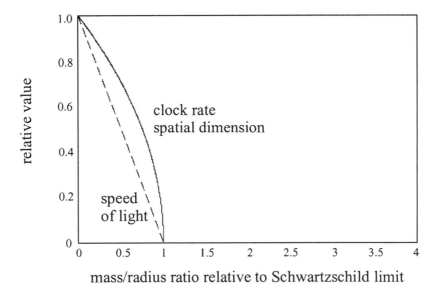

Figure 6-7: Variation of speed of light, clock rate, and a spatial dimension for Einstein theory versus the mass/radius ratio at the surface of a star relative to the Schwartzschild limit

Computer Solutions of General Relativity

Einstein could apply general relativity only to very simple cases. With more complicated applications, the equations of general relativity yield millions of terms, and so cannot be solved analytically. In the 1960's, a decade after Einstein's death, powerful computers became available that could apply general relativity to complicated physical models. Since then, hundreds of scientists throughout the world have devoted their careers to computer solutions of the Einstein equations. This effort has been directed almost exclusively to cosmology, because that is about the only area that can use this expertise.

Chapter 7

Einstein's Rejection of Singularities

As was explained in Chapter 5 (p. 96), Stephen Hawking and Roger Penrose published a paper in 1970, which claimed to have proven that "if Einstein mathematics were correct, a singularity had to result from a black hole" and "the universe must have started with a Big Bang explosion out of a singularity. The equations do not allow an alternative".

Nevertheless, in 1939, Einstein absolutely rejected the *black hole* singularity, and in 1945 he absolutely rejected the *Big Bang* singularity. This chapter examines these issues. We start with the black hole singularity.

The Black Hole Singularity

Theory of the Neutron Star

As explained by Goldsmith [25] (p. 193), the theory of neutron stars began in earnest under the guidance of Robert Oppenheimer, who later directed the Manhattan atomic bomb project. In 1939 he published a paper on neutron stars with the assistance of a graduate student, George Volkoff. This *Physical Review* paper was entitled "On Massive Neutron Stars". This gave the first firm prediction of how massive a star must be to become a neutron star.

The density of a neutron star is 200 million metric tons per cubic centimeter. A neutron star with 9.6 times the mass of our sun would have a mass/radius ratio equal to the Schwartzschild limit (240,000 times the mass/radius ratio of our sun.)

As was explained in Chapter 4, when a star with more than 8 times the mass of our sun depletes its nuclear fusion fuel, it collapses catastrophically to form a neutron star. The gravitational energy released by the catastrophic collapse blows the outer portion of the star apart in a supernova explosion. Because much of the star is blown away, the neutron star that remains has appreciably less mass than the original star. On the other hand, there are stars with masses up to 100 times that of our sun, and some of these should produce neutron stars that are much larger than 9.6 times the mass of our sun. The mass/radius ratio of such a neutron star should exceed the Schwartzschild limit.

This reasoning indicated to Oppenheimer that the Schwartzschild limit was not a theoretical issue without practical meaning. Our universe should have neutron stars with mass/radius ratios that exceed the Schwartzschild limit. What should happen to such a neutron star? Oppenheimer addressed this issue with a second 1939 *Physical Review* paper. This September 1939 paper, with graduate student H. Snyder, was titled "On Continued Gravitational Contraction". [12]

The Schwartzschild analysis assumes that the star has a constant size. Oppenheimer and Snyder found that they could achieve a solution from the Einstein equations, when the *Schwartzschild limit* is exceeded, provided that they assumed that the diameter of the star decreases with time. As the star shrinks, its mass/radius ratio increases, and so the star should shrink further. The paper concluded with:

"When all thermonuclear sources of energy are exhausted, a sufficiently heavy star will collapse. Unless fission due to rotation, the radiation of mass, or the blowing off of mass by radiation, reduce the star's mass to the order of that of the sun, this contraction will continue indefinitely."

In other words, this statement claimed that (according to the Einstein theory) a very massive star must collapse when the *Schwartzschild limit* is exceeded, and must shrink "indefinitely" until it becomes a *"singularity"* having an *infinite density of matter*.

The next month Einstein responded to this paper with an extensive analysis in *Annals of Mathematics* [10], but the Einstein paper politely did not specifically refer to the Oppenheimer-Snyder article. Einstein concluded with

"The essential result of this investigation is a clear understanding as to why the 'Schwartzschild singularities' do not exist in physical reality. Although the theory here treats clusters whose particles move along circular paths, it does not seem to be subject to reasonable doubt that more general cases will have analogous results. The 'Schwartzschild singularity' does not appear for the reason that matter cannot be concentrated arbitrarily. And this is

due to the fact that otherwise the constituting particles would reach the velocity of light."

A few years later, Oppenheimer became the manager of the Manhattan atomic bomb project and never pursued this issue further. Neither did any other scientist while Einstein was alive.

In his rebuttal of the Oppenheimer-Snyder paper, Einstein insisted that the *singularity* condition derived from his theory is non-physical. The proposition that a star could contract indefinitely to form a *singularity* of infinite mass density severely violates our laws of physics. **Einstein never accepted the physically impossible singularity**

The Black Hole Concept

Although the concept that Oppenheimer and Snyder presented was not pursued officially while Einstein was alive, it did not die. When the mass-to-radius ratio of a star exceeds the Schwartzschild limit, the theory indicates that the star should be surrounded by a spherical surface called an *"event horizon"*, over which the speed of light is zero. On the event horizon sphere, the ratio of mass-to-radius is equal to the Schwartzschild limit. Light theoretically cannot escape from within the event horizon sphere, and so the star became known as a *"black hole"*.

The *black hole* became a popular theme for science fiction, and the public became well aware of the concept. There were many descriptions of the process of *"falling into a black hole"*. Despite the wide popular knowledge of the black hole, few people understand what must happen to the black-hole star that lies within the event horizon sphere. They do not realize that the black hole star must theoretically collapse "indefinitely" until it shrinks to form a "singularity" having zero diameter and an infinite density of matter.

The diameter of the event-horizon sphere is equal to 6 kilometers multiplied by the mass of the star divided by the mass of the sun. The event-horizon sphere is all that one can theoretically observe from the outside world when looking at a black hole. Light cannot escape from within the event-horizon sphere.

Some of the science-fiction writers who considered the black hole have recognized that very strange conditions should exist inside the event horizon. Since light can only fall toward the center of a black hole, two adjacent atoms of a body inside the event horizon could not interact in a physically meaningful manner. Energy could be transmitted from the upper atom to the lower atom, but not in the reverse direction. How then can real matter exist inside the event horizon?

The answer is that physical matter can only exist momentarily inside the event horizon of a black hole. If an object falls through the event horizon, it theoretically must fall "indefinitely" toward the center of the black hole until it is squeezed to zero size and infinite density.

Physical Evidence for Black Holes

About a decade after Einstein's death, powerful computers became widely available, and many scientists began to apply them to the Einstein theory. These computer studies supported the conclusion of Oppenheimer and Snyder given on page 125. Contrary to Einstein's objection, the computer studies found that the "Schwartzschild singularity", which is associated with the "black hole", is a required prediction of the Einstein equations. Consequently the black hole concept became generally accepted by scientists.

In recent years, many astronomers have made observations that are considered to be evidence of black holes. But how does one prove the existence of a black hole? A black hole does not emit any radiation to indicate its presence.

Astronomers are observing the gravitational effects of massive, compact bodies. According to the observational evidence, these so-called black-hole bodies could also be massive neutron stars. However, the solutions of the Einstein equations have predicted that the Einstein theory cannot allow a massive neutron star; such a star would have to collapse to form a black hole. ***Consequently, astronomers reject the neutron star possibility and insist that they are observing black holes, even though the black hole singularity violates our laws of physics and was absolutely rejected by Einstein.***

This point is illustrated by Dickinson [28] (p. 89). The Cygnus X1 star system contains a blue giant star, 27 times the sun's mass, orbiting with an invisible star, 15 times the sun's mass. The distance between the two stars is 1/5 of the earth's orbital radius. Astronomers have concluded that the invisible star is too massive to be a neutron star, and so must be a black hole.

The Big Bang Singularity

Einstein recognized, as did Gamow, that nuclear matter has the greatest possible density of matter. The concept that our universe could be compressed to a "singularity" density that is many, many times greater than the density of nuclear matter strongly conflicts with physical evidence. Einstein never accepted the singularity concept.

In 1945, Einstein recognized that his theory implies a singularity at the birth of the universe. He flatly rejected this interpretation of his theory with the following (Ref. [5], p. 129):

"Theoretical doubts [concerning the creation of the universe] are based on the fact that [at the] beginning of the expansion, the metric becomes singular and the density becomes infinite. . . In reality, space will probably be of a uniform character, and the present [relativity] theory will be valid only as a limiting case. . . **One may not therefore assume the validity of the equations for very high density of field and of matter, and one may not conclude that the 'beginning of the expansion' must mean a singularity in the mathematical sense.** *All we have to realize is that the equations may not be continued over such regions."*

In this quotation, Einstein stated that his theory could not be used to justify a physical singularity, because his equations would not apply under conditions of extreme density of matter. The thinking of Albert Einstein on such issues is discussed in a recent biography of Einstein, originally written in German by Folsing [23], which states (p. 381):

"Some of Einstein's admirers were tempted to see the general theory of relativity as a triumph of speculation over empiricism. This kind of misunderstanding made Einstein 'downright angry' [who said] 'This development teaches us something entirely different, indeed almost the opposite, namely that a theory, in order to merit confidence, must be based on generalizeable facts'. . . . To Einstein, facts were not only the starting point of his theory but also the keynote of any test of it."

Einstein had extensive experience with physical experiments and was absolutely committed to the principle that theory must agree with physical evidence. The claim that the Einstein theory could be used to make singularity predictions that grossly conflict with experimental evidence is a drastic violation of Einstein's scientific philosophy. The landmark papers published by Einstein in 1905 (discussed on page 108) illustrate the solid experimental foundation of Einstein's research.

Are the Einstein Equations Correct?

The Einstein general relativity equations are explained in Appendix E, Section E.2. That appendix shows that the Einstein general theory is specified by a tensor formula called the Einstein gravitational field equation, which represents 10 independent equations. This tensor formula is extremely complicated, because it is expressed in terms of the tensors of the complex Riemann-Ricci mathematics of curved space.

The justifications for the bizarre predictions associated with the black hole and the Big Bang singularities are based solely on the equations of the Einstein general theory of relativity. We have seen that Einstein flatly rejected these singularity predictions. An obvious explanation for this contradiction is that there is something wrong with the Einstein gravitational field equation.

The derivation of the Einstein gravitational field equation is discussed by John A. Peacock in his lengthy scientific book, *Cosmological Physics* [22]. This book describes theoretical Big Bang research and includes a detailed discussion of relativity theory.

Peacock states that the Einstein gravitational field equation "cannot be derived in any rigorous sense; all that can be done is to follow Einstein and start by thinking about the simplest form such an equation might take." [22] (p. 19)

On pp. 26-27, Peacock [22] discusses possible *"alternative theories of gravity"*. He asks, *"How certain can we be that Einstein's theory of gravitation is correct?"* He notes that the Einstein theory does not apply in sub-atomic scales, and goes on to say, *"Apart from this restriction, there are no obvious areas of incompleteness. . . Nevertheless, . . . it is possible that more accurate experiments will yield discrepancies. Over the years this possibility has motivated many suggestions of alternatives to general relativity."*

Peacock is a strong supporter of the Big Bang theory, which is solidly tied to the Einstein gravitational field equation. Nevertheless, even Peacock admits that the Einstein gravitational field equation was obtained in an intuitive manner, and that responsible scientists have seriously considered a number of alternatives to that equation.

The Yilmaz Refinement of the Einstein Theory

The Yilmaz theory of gravity will be explained in Chapter 8. The Yilmaz theory applies the principles of the Einstein general theory of relativity and so is a refinement of the Einstein theory. Yilmaz discovered the key that allowed him to calculate an exact solution to the principles of general relativity that Einstein had established. From this exact solution, Yilmaz derived in a rigorous manner a different gravitational field equation, which is the foundation for the Yilmaz theory of gravity.

If the Yilmaz theory is correct, then the Einstein gravitational field equation is wrong. However, the basic Einstein general theory of relativity is still correct, because the Yilmaz theory is a refinement of the Einstein theory.

As we will see, the Yilmaz theory does not allow a singularity. The Yilmaz theory does not predict either a black hole singularity or a Big Bang singularity. Therefore the Yilmaz refinement of the Einstein theory agrees that Einstein was correct when he absolutely rejected the singularity concept.

Chapter 8

The Yilmaz Theory of Gravity

Basis for the Yilmaz Theory

In the early 1950's, Huseyin Yilmaz was studying general relativity in his PhD research at the Massachusetts Institute of Technology. He examined Einstein's approximate calculation of the effect of gravity on wavelength, which was discussed in Chapter 6. Yilmaz discovered that he could implement this analysis exactly. This yielded an exact solution to the principles of general relativity. The resultant Yilmaz theory of gravity was published in the prestigious *Physical Review* in 1958. [Y1] The following summarizes the Yilmaz analysis.

Exact Effect of Gravity on a Time Measurement

As was shown in Chapter 6, Einstein proved that when light rises against gravity it experiences a redshift ($\Delta L/L$) approximately equal to (gh/c^2). From this result Einstein obtained an approximate formula for the effect of gravity on a time measurement.

Yilmaz discovered an exact solution to this approximate analysis. As explained in Appendix E, Section E.1, Yilmaz derived an exact formula for the redshift produced by gravity. From this he obtained an exact formula for the relativistic effect of gravity (and acceleration) on a *time measurement*.

Exact Effect of Gravity on a Distance Measurement

To generalize his result, Yilmaz postulated that ***the speed of light measured locally in a gravitational field is independent of direction.*** By combining this postulate with his exact formula for the relativistic effect of gravity on a *time measurement*, Yilmaz derived an exact formula for the relativistic effect of gravity on a *distance measurement*.

Yilmaz Gravitational Field Equation

From his exact formulas for the relativistic effect of gravity on time and distance measurements, Yilmaz calculated in a rigorous manner the *gravitational field equation* that specifies his theory.

The Yilmaz gravitational field equation has an additional tensor that characterizes the force and energy of the gravitational field. This tensor is not included in the Einstein theory. As explained in Appendix E, Section E.3, this additional tensor is essential to allow a relativistic gravitational theory to yield a rigorous solution.

The Yilmaz theory is very much easier to use than the Einstein theory. The gravitational field equation of the Yilmaz theory never needs to be solved, because Yilmaz has a general solution to this equation.

Time-Varying Solution to the Yilmaz Theory

The solution to the Yilmaz theory described above is a "static" solution, which theoretically applies exactly only when the gravitational field does not change with time. Yilmaz has also derived a much more complicated time-varying solution to his theory, which does not have this limitation.

This time-varying solution proves that the simple static Yilmaz solution gives a very accurate result if the gravitational field varies slowly relative to the speed of light, a condition satisfied in essentially all practical applications. About the only case where the general time-varying solution is needed is in the analysis of gravitational waves.

Lack of Singularities in the Yilmaz Theory

The Einstein gravitational field equation predicts black hole and Big Bang singularities, but the Yilmaz theory does not allow a singularity.

The Black Hole Singularity

The lack of a black-hole singularity in the Yilmaz theory is illustrated in Figs. 8-1 and 8-2. These figures compare the speed of light, clock rate, and spatial dimension predicted by the Einstein and Yilmaz theories for stars of different mass/radius ratios. The solid curves are the predictions of the Yilmaz theory and the dashed curves are the predictions of the Einstein theory that were given in Fig. 6-7. These plots are expressed in terms of the mass/radius ratio relative to the Schwartzschild limit. (The Schwartzschild limit is 240,000 times the mass/radius ratio at the surface of our sun.) Figure 8-1 compares the speed of light predictions of the Yilmaz and Einstein theories, and Fig. 8-2 compares the reductions of clock rates and spatial dimensions predicted by the two theories. These plots are derived in *Story* [2], Chapter 10 and Appendix E.

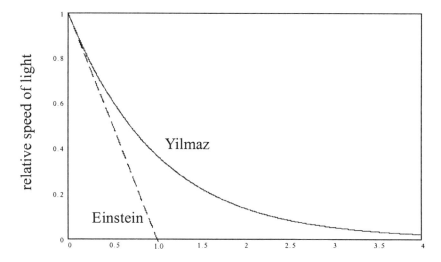

mass/radius ratio relative to Schwartzschild limit

Figure 8-1: Relative speed of light for Yilmaz theory (solid curve) and Einstein theory (dashed curve) versus the mass/radius ratio at the surface of a star

For the Einstein theory, the plots all go to zero at the Schwartzschild limit, but for the Yilmaz theory the plots never go to zero. Consequently, the Yilmaz theory does not allow a *black hole* singularity.

These plots in Figs. 8-1 and 8-2 demonstrate that the physically impossible singularity condition associated with the black hole is merely a mathematical defect in the Einstein gravitational field equation. In agreement with Einstein's statement on page 128, the Yilmaz theory demonstrates that "Schwartzschild [black-hole] singularities do not exist in physical reality".

All of the evidence that astronomers have found for black holes can also be interpreted as evidence for massive neutron stars. Astronomers reject the neutron star possibility because the Einstein

gravitational field equation does not allow a massive neutron star. However, the Yilmaz refinement of the Einstein theory shows that neutron stars can have extremely high mass.

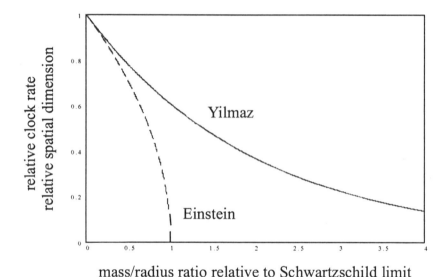

Figure 8-2: Reduction of clock rate and spatial dimension predicted by Yilmaz theory (solid curve) compared with Einstein theory (dashed curve) versus the mass/radius ratio at the surface of the star

For small values of the mass/radius ratio, the plots of the Yilmaz theory closely approximate those of the Einstein theory. For experiments performed within our solar system, the mass/radius ratios are very small, and so the Einstein and Yilmaz theories yield essentially the same results. Therefore, all of the experiments that have verified the Einstein theory are also consistent with the Yilmaz theory.

The Big Bang Singularity

Chapter 10 will describe the cosmology theory that is predicted by the Yilmaz theory of gravity. The Yilmaz theory predicts a Steady State theory of the universe, rather than a Big Bang theory. There is no singularity condition associated with the Yilmaz cosmology theory. Hence the Yilmaz theory does not allow either a Big Bang singularity or a black hole singularity.

Consistency with Quantum Mechanics

After presenting his general theory of relativity, Einstein did little with this theory. He devoted the rest of his life primarily to the search for a *Unified Field Theory*. This would have combined into a single theory the concepts of gravitational fields, electromagnetic fields, and atomic nuclear fields. He never succeeded, although he struggled with this task until his last days. An important reason for his failure is that the Einstein gravitational field equation conflicts with quantum mechanics.

Yilmaz has proven that his gravitational field equation is consistent with quantum mechanics. [Y11] This finding opens great possibilities for relating relativistic gravitational theory to quantum field theory, and should achieve Einstein's elusive goal of a *Unified Field Theory*.

String Theory

In the last few years, the esoteric mathematics of *String Theory* have been widely publicized to the general public. This theory postulates that all matter consists of infinitesimal oscillating elements that vibrate like the strings of a violin. These vibrating "strings" are smaller than one billionth of the size of a proton, so small they cannot be observed directly. The strings oscillate in 11 independent dimensions.

Einstein's theory predicts that reality is defined in four independent dimensions: three spatial dimensions plus time. String theory postulates that, in the infinitesimal world of its vibrating strings, the space has seven extra dimensions that cannot be observed in our normal world. The vibrating strings of String theory are so small they are undetectable, and vibrate in 11 independent dimensions, 7 of which are unobservable.

What is the evidence to support these extreme postulates? These postulates evolved from a complex mathematical theory that succeeded in reconciling (in a mathematical sense) the Einstein gravitational field equation with the principles of quantum mechanics. The purpose is to satisfy Einstein's goal for a *Unified Field Theory*, which reconciles gravitational force with electromagnetic and nuclear forces. The *string theorists* claim that they are developing a *"Theory of Everything"*.

The Yilmaz Alternative to String Theory

The Yilmaz theory provides a very simple alternative to String Theory. The Yilmaz theory is a refinement of the Einstein theory that satisfies the principles established by Einstein in developing his general theory of relativity. Since the Yilmaz theory is consistent with quantum mechanics [Y11], the Yilmaz theory directly reconciles relativistic gravitational force with electromagnetic and nuclear forces. Consequently there is no need for the bizarre postulates of *String Theory*. The Yilmaz theory already provides a *"Theory of Everything"*, and should achieve Einstein's goal of a *Unified Field Theory*.

Variation of Speed of Light with Direction

For tests performed within our solar system, the Einstein and Yilmaz theories yield essentially the same results. Consequently all tests that have verified the Einstein theory are consistent with the Yilmaz theory.

However, there is one very sensitive test, which has been partially implemented, that could distinguish between the Einstein and Yilmaz theories. A fundamental principle of the Yilmaz theory is that the speed of light measured locally in a gravitational field is the same in all directions. The Einstein theory predicts it is different.

It is very difficult to measure with sufficient accuracy the variation of the speed of light with direction, because the test cannot use two-way transmission of light. As explained by Prof. Carroll O. Alley [Y10], preliminary experiments were performed under his guidance to implement this measurement. This involved the transportation of an atomic clock, back and forth between the U.S. Naval Observatory and the NASA Goddard Optical Research Facility, which are 21.5 km apart. The transported clock was used to synchronize atomic clocks in the two locations.

The initial data were very promising. However, government funding was cancelled before measurements could be made with the required accuracy to obtain definitive results. Why was the modest funding for this very important experiment cancelled?

This experiment was discussed by Ivars Peterson [42] in a 1994 *Science News* article: *"A New Gravity: Challenging Einstein's general theory of relativity"*. Further discussion of this issue is given in Appendix E, Section E.3.

Uniqueness of the Yilmaz Theory

A confusing aspect of the Einstein theory is that it can yield multiple solutions for the same physical model. Constraints that are somewhat arbitrary must be included in a general relativity analysis to achieve an answer. This problem is widely recognized by those performing general relativity studies.

In contrast, the Yilmaz theory has a definite solution, and can yield only one answer for a particular physical model. The Yilmaz theory incorporates general constraints, which assure that this condition is always satisfied.

Scientists who are familiar with the arbitrary adjustable parameters of the Einstein theory may find it difficult to recognize that the Yilmaz theory does not have this property. A prediction made by the Yilmaz theory depends only on the characteristics of the physical model on which it is based. It is not affected by any assumption made by the individual who is applying the Yilmaz theory.

Inter-Stellar Space Travel

To allow a space ship to travel the enormous distances between stars in a reasonable time, science-fiction writers frequently use the concept of the *"wormhole"*, which is a *short-cut* across space. The *wormhole* is related to the *black hole*, and has been justified by the peculiar physical conditions predicted inside a black hole. Big Bang cosmologists do not directly support the wormhole concept, but often take the position that they cannot prove it is impossible. (See Dickinson [28] p. 91.)

In agreement with Einstein, the Yilmaz theory demonstrates that the black hole is physically impossible, and so the wormhole is also physically impossible.

Will humans ever achieve inter-stellar space travel? If they do, there will be no short-cuts. Our present maximum velocity of interplanetary space vehicles is about 12 km/sec. Vehicle velocity would have to increase by 25,000 to reach the speed of light. However, that velocity cannot be achieved, because the mass of the space vehicle would become infinite, as was shown in Fig. 6-4 on page 114.

If the space ship velocity reaches 86.6 percent of the speed of light, the mass of the space ship would double. To reach this speed, the momentum (mass times velocity) must be 43,000 times that of present space vehicles. At this speed, it would take *10 years* to make a round trip to the nearest stars of interest, the Alpha Centauri binary pair of stars 4.3 light years away, which are about the size of our sun. However, it is unlikely that a binary star can have an earth-like planet. A much longer trip is probably needed to reach an earth-like planet that can support life.

Chapter 9

Lack of Objectivity in Astronomy

The Bandwagon Mentality in Astronomical Research

Astronomers today present their findings in television, magazines, and books with absolute confidence. They refer to the Big Bang event as if it were an established fact. Other astronomical concepts are proclaimed with equal assurance. From this confidence, one might think that these conclusions are based on overwhelming evidence, but the opposite is true. The confidence is covering up great gaps and contradictions in the evidence.

Astronomy is controlled by a bandwagon mentality that is stifling dissent. Without open scientific debate there can be no true science. Concepts that are inconsistent with accepted dogma are rejected as being irresponsible.

This chapter addresses the lack of objectivity in astronomical research. We start with an editorial by the noted astrophysicist Geoffrey Burbidge in the February 1992 *Scientific American*, which strongly criticized the bandwagon mentality of Big Bang astronomy.

Next we consider the case of the noted astronomer Halton Arp, who was denied access to the Palomar and Mount Wilson Observatories in 1984, after 30 years of distinguished research, because he had found strong astronomical evidence that contradicted accepted dogma concerning the quasar.

Finally, we examine the case of the plasma physicists, supported by Nobel laureate Hannes Alfven. Astrophysical journals refuse to accept the research of plasma physicists, because it is inconsistent with prevailing astronomical dogma. This case was presented in a best selling book by Eric Lerner.[16]

The Editorial of Geoffrey Burbidge

The case against the Big Bang was expressed eloquently by Professor Geoffrey Burbidge in an editorial article of the February 1992 *Scientific American* [34]. Professor Burbidge is the former director of the Kitt Peak National Observatory, and is presently Professor of Astrophysics at the University of California in San Diego. Prof. Burbidge began his editorial with:

> *"Big bang cosmology is probably as widely believed as has been any theory of the universe in the history of Western civilization. It rests, however, on many untested, and in some cases untestable, assumptions. Indeed, big bang cosmology has become a bandwagon of thought that reflects faith as much as objective truth."*

He went on to say

> *"Younger cosmologists are even more intolerant of departures from the big bang faith than their more senior colleagues are. Worst of all, astronomical textbooks no longer treat cosmology as an open subject. Instead the authors take the attitude that the correct theory has been found."*

Burbidge then explained the basic reasons for the bandwagon mentality:

> *"Powerful mechanisms encourage this conformity. Scientific advances depend on the availability of funding, equipment, and journals in which to publish. Access to these resources is granted through a peer review process. Those of us who have been around long know that peer review and the refereeing of papers have become a form of censorship. It is extraordinarily difficult to get financial support or viewing time on a telescope unless one writes a proposal that follows the party line."*

> *"A few years back, Halton C. Arp was denied telescope time at Mount Wilson and Palomar Observatories because his observing program had found, and continued to find, evidence contrary to standard cosmology."*

> *"Unorthodox papers often are denied publication for years or are blocked by referees. The same attitude applies to academic*

positions. I would wager that no young researcher would be willing to jeopardize his or her scientific career by writing an essay such as this."

Burbidge concluded with the following:

"The big bang ultimately reflects some cosmologist's search for a creation and a beginning. This search properly lies in the realm of metaphysics, not science."

Halton Arp's Quasar Discoveries

Discovery of the Quasar

After World War II, a great many radio antenna systems were constructed to act as "radio telescopes" that detected the radio emissions from the heavens. These investigations led to the discovery of a strange new type of star, which was called a *quasi-stellar radio object*, and was given the acronym *quasar*.

The resolution of radio telescopes is poor because of the long wavelengths of the signals. Much higher resolution was achieved for quasars eclipsed by the moon. The instant at which a quasar disappeared behind the moon was detected. The radio signal was compared with a nearby optical star image eclipsed by the moon at the same instant. This allowed a radio signal to be related accurately to an optical star image.

Astronomers began to study the optical spectra of these radio objects, and found them to be peculiar. The spectra did not look like the spectra of known elements. Finally in Feb. 1963, astronomers Jesse L. Greenstein and Maarten Schmidt jointly discovered that they could explain the quasar spectra by assuming that quasars have extremely large spectral redshifts. [40]

It was soon discovered that not all of these strange stars with extreme redshifts are radio sources. Consequently their name was changed from *quasi-stellar radio objects* to *quasi-stellar objects*, but the acronym *quasar* was retained. They are also called *QSO's*.

The anomalous redshift of the quasar is a serious enigma. If this redshift is a Doppler effect caused by velocity, the quasars must be receding from us at extremely high velocities. This in turn requires them to be billions of light years away. For quasars to be at such vast distances, they must radiate enormous power for us to see them.

For example, one of the first two quasars to be studied was quasar

3C48, which is examined in Appendix B of *Story* [2]. It has a redshift of 0.367, which means that the wavelength shift is 36.7 percent of the normal wavelength. If this is a Doppler effect due to velocity, the quasar must be moving away from us at approximately 36.7 percent of the speed of light, which comes to 110,000 km/sec.

Greenstein and Schmidt assumed a Hubble constant of 31 km/sec per million light years of galaxy distance, which was the generally accepted value for the Hubble constant at that time. From this they estimated the distance of quasar 3C48 to be 3.6 billion light years. At that distance this quasar would have to radiate the power of 1000 billion suns to produce the stellar radiation that they received. This radiation is 100 times the total power radiated by our enormous Milky Way Galaxy.

Then they discovered that the power of 3C48 varied by 40 percent over a period of 600 days. This rapid variation indicated that 40 percent of this enormous quasar power must be generated within a volume with a thickness of only 600 light days or 1.6 light years. ***Within a volume that is only 1.6 light years thick, quasar 3C48 is presumably generating 40 times the total power of our enormous Milky Way galaxy, which is 100,000 light years in diameter.*** [40]

Later studies have found quasars with much larger redshifts. Studies have also shown that some quasars vary in power over much shorter periods. Quasars have been discovered that vary by a factor of 2 in power within several hours, and so can be no larger than our solar system. [41]

Quasar Observations of Halton Arp

Because of the unbelievable properties attributed to quasars, the noted astronomer, Halton Arp, suspected that they might be much closer than was being assumed. He started making observations of quasars to obtain direct astronomical estimates of their distances.

Arp found many quasars with images very close to galaxies having much smaller redshifts. If the image of one quasar is very close to that of a galaxy, this might be a chance relationship. The quasar might be billions of light years beyond the galaxy. However, if two or more quasars appear to be close to a galaxy, the probability of a chance relationship is remote. We ask, "What is the probability that the images of two or more quasars would fall this close to an arbitrary direction in space?" Probability considerations indicate that it is highly unlikely that the quasars are not physically close to the associated galaxy.

Arp photographed three quasars that appear to be in the outer fringe of galaxy NGC 3842. It is theoretically possible that this is a chance relationship; that the quasars actually lay far beyond the galaxy. However, the probability that three quasar images would fall this close to an arbitrary direction in space is about one in a million.

Arp found several cases like this. The possibility is essentially zero that all of these observations could be accidental, which relate quasars to galaxies having much smaller redshifts.

In Halton Arp's study of quasars, he also found many cases of filament structures that directly connect quasars to galaxies having much smaller redshifts. These observations suggest that the quasar was ejected from the associated galaxy by a supernova explosion. There were several cases of a filament connecting a quasar to a galaxy, and another opposing filament on the other side of the galaxy. This suggests that the opposing filament is the reaction from a supernova explosion that ejected the quasar from the galaxy.

The astronomical community opposed the quasar observations by Arp, because they did not agree with the accepted dogma that quasars are billions of light years away. He found it very difficult to get his findings published. Some journals rejected his work, and papers were often held up for years by referees. Finally in 1984, the committee that controls observation time at Palomar and Mount Wilson Observatories refused to allow Arp to use these facilities.

Halton Arp had performed distinguished research at Palomar and Mount Wilson Observatories since he received his PhD degree in 1953. He was president of the Astronomical Society of the Pacific from 1980 to 1983, and received awards from the American Astronomical Society, the American Association for the Advancement of Science, and the Alexander von Humbolt Senior Scientist Award. After being denied research facilities in California, he was forced to move to Germany to continue his career, where he joined the Max Planck Institute for Physics and Astrophysics in Munich.

Arp presented his quasar observations up to the time of his expulsion in his 1987 book, *Quasars, Redshifts, and Controversies* [14]. In his later 1998 book, *Seeing Red* [15], he also includes the extensive observations on quasars that he has made since he was forced to move to Germany. Arp's latest book gives overwhelming evidence that quasars are very much closer than is generally assumed. Nevertheless, officials in astronomy completely ignore this evidence, and continue to insist that the conventional quasar dogma is correct.

The traumatic experience of Halton Arp was summarized as follows by Fred Hoyle, Geoffrey Burbidge, and Jayant V. Narlikar [19] (p. 134)

9. Lack of Objectivity in Astronomy   143

in their (year 2000) book, *A Different Approach to Cosmology*:

> *"Arp's own colleagues at the Mount Wilson and Palomar Observatories . . . recommended to the directors of the two observatories that his observational program should be stopped, i.e., that he should not be given observing time on the [telescopes in these observatories] to carry on with this program. Despite his protests, the recommendation was implemented, and after his appeals to the trustees of the Carnegie Institution were turned down, he took early retirement and moved to Germany where he now resides, working at the Max-Planck-Institut fur Physik und Astrophysik in Munich. . . . Thus Arp was the subject of one of the most clear cut and successful attempts in modern times to block research which it was felt, correctly, would be revolutionary in its impact if it were to succeed."*

What Is a Quasar?

The observations of Halton Arp give strong evidence that the extreme redshift of the quasar is not produced by velocity. Arp says that a quasar has *"intrinsic redshift'*, which is unrelated to velocity. What is the cause of intrinsic redshift?

Arp has shown that galaxies as well as quasars can have intrinsic redshift. He has photographed many cases of a galaxy that is directly connected by a filament structure to a larger galaxy having much lower redshift. The evidence often suggests that the smaller galaxy was ejected by the larger galaxy.

A promising explanation for intrinsic redshift is a theory proposed by Paul Marmet. [37] As was shown in Chapter 5 (pp. 90-91), Marmet has proposed his redshift effect to explain the Hubble expansion of the universe. He believes that the universe is not actually expanding, and that an effect other than velocity is producing the galaxy redshift, which Hubble interpreted as an expansion of the universe. Marmet has shown that a cloud of hydrogen gas can produce a strong redshift.

It was shown in Chapter 5 that the Marmet effect is not sufficient to explain the Hubble expansion of the universe. However, it gives a promising explanation for the intrinsic redshift of the quasar. As explained in *Story* [2], Appendix A, a hydrogen cloud surrounding a quasar with a density of 100,000 hydrogen atoms per cubic centimeter would produce a redshift of 0.2 for every 1000 light years of cloud thickness. Many gaseous nebulae have similar values of hydrogen gas density. This redshift effect appears to be sufficient to explain the intrinsic redshifts of quasars and galaxies.

As explained in Chapter 8, relativistic effects produced by gravity can also produce a strong redshift. The gravitational redshift predicted by the Einstein theory is limited, but the Yilmaz theory can predict a very large gravitational redshift. The star would have to be very dense. Consequently, gravitational redshift cannot explain the intrinsic redshift of galaxies, but may explain the intrinsic redshifts of quasars.

Here are two promising explanations for intrinsic redshift. They are much more consistent with experimental evidence than the bizarre quasar concepts presently endorsed by astronomers.

Eric Lerner and Nobel Laureate Hannes Alfven

In 1991, Eric Lerner wrote the book, *The Big Bang Never Happened* [16]. He was strongly supported in this book by Nobel laureate Hannes Alfven (1908-1995), who was the founder of modern plasma physics. The stimulus for the book was the strong opposition that Hannes Alfven and other plasma physicists had received in their attempts to publish papers in astronomical journals that relate plasma physics to cosmology.

The Cosmology Theories of Hannes Alfven

Chapter 4 (p. 66) described the Alfven theory of creation of our solar system. This theory showed how plasma electric currents could have transferred angular momentum from the sun to its solar system when our sun was being formed.

Alfven has also concluded that plasma electric currents cause the rotation of galaxies. In laboratory experiments, he demonstrated that instabilities cause plasma electric currents to twist around one another like the strands of a rope. This effect can theoretically cause a galaxy to rotate. Rotation of a galaxy in turn can cause the rotation of a star when a huge cloud of hydrogen gas collapses to form the star.

This effect is similar to the whirlpool that is produced when water pours down a sink drain. When the water starts to flow, it has a slight rotation because the sink and drain are not completely symmetric. As the water flows, instability in the flow process quickly amplifies this slight rotation, until a strong whirlpool is formed. (Contrary to a popular myth, the rotation of the earth has negligible effect on the rotation of the whirlpool.)

Just as rotational instability in the flow of water produces a whirlpool, so the rotational instability in the flow of plasma electric currents can theoretically cause a galaxy to rotate.

Chapter 4 showed (p. 93) that nearly all galaxies are organized into curling filament structures that extend for hundreds of millions of light years. Alfven has shown theoretically that the instability produced by plasma electric currents can create the curling filament structures into which galaxies are organized.

Much more information is given by Lerner [16] concerning the tremendous cosmological implications of plasma physics, which has been pioneered by Nobel laureate Hannes Alfven.

Mythological Philosophy of Big Bang Research

Big Bang cosmologists have ignored or dismissed plasma theory, and few have even bothered to read about it. The well-known cosmologist, James Peebles (called the "father of modern cosmology" by *Scientific American*) stated that Alfven's ideas are *"just silly"*. His colleague at Princeton, Jeremiah Ostriker, commented, *"There is no observational evidence that I know of that indicates electric and magnetic forces are important on cosmological scales."*

Alfven, as well as lesser-known plasma physicists, have repeatedly had their papers rejected by astrophysical journals because they contradict Big Bang wisdom. Alfven commented, *"I think the Catholic Church was blamed too much for the case of Galileo — he was just a victim of peer review"*.

As evidence against the Big Bang theory has mounted, Big Bang cosmologists have shrugged it off, and have proceeded to devise more and more elaborate ad-hoc theories to bypass the evidence. Joseph Silk, who has written three books on the Big Bang, stated flatly

> *"It is impossible that the big bang is wrong. Perhaps we'll have to make it more complicated to cover the observations, but it is hard to think of what observations could refute the theory itself."*

Lerner [16] (page 54) responded to this with the following, which must also have reflected the convictions of Nobel laureate Alfven, who helped Lerner greatly in his book:

> *"This attitude is not at all typical of the rest of science, or even of the rest of physics. In other branches of physics, the multiplication of unsupported entities to cover up a theory's failure would not be*

tolerated. The ability of a scientific theory to be refuted is the key criterion that distinguishes a science from metaphysics. If a theory cannot be refuted, if there are no observations that could disprove it, then nothing can prove it — it cannot predict anything; it is a worthless myth."

Lerner (p. 56) quotes the following words by Alfven on the myth issue:

"The cosmology of today is based on the same mythological views as that of the medieval astronomers, not on the scientific traditions of Kepler and Galileo."

Big Bang theorists treat the Einstein gravitational field equation as their ultimate truth. As reported by Lerner (p. 163), cosmology theorist George Field stated his Big Bang philosophy as:

"I believe the best method is to start with exact theories, like Einstein's, and derive results from them."

This philosophy of modern cosmology sharply contrasts with legitimate science, which demands consistency between theory and observation. It is the philosophy of mythology, not of science. Lerner (p. 162) states:

"Entire careers in cosmology have now been built on theories that have never been subject to observational tests, or have failed such tests and have been retained nonetheless."

As Lerner (p. 127) points out, the mythology of modern cosmology is based on the "myth of Einstein". He explains Alfven's concepts with:

"It is quite ironic that the triumph of science [from relativity theory] led to the resurgence of myth. The most unfortunate effect of the Einstein myth is the enshrinement of the belief, rejected for four hundred years, that science is incomprehensible, that only an initiated priesthood can fathom its mysteries."

Lerner quotes the following words of Alfven:

"The people were told that the true nature of the physical world could not be understood except by Einstein and a few other geniuses who were able to think in four dimensions. Science was something to

9. Lack of Objectivity in Astronomy   147

believe in, not something that should be understood. Soon the best sellers among the popular science books became those that presented scientific results as insults to common sense. One of the consequences was that the boundary between science and pseudo-science began to be erased. To most people, it was increasingly difficult to find any difference between science and science fiction."

Since the start of the wide acceptance of the Big Bang theory about 1965, cosmology theorists have been struggling with its many serious conflicts with observational evidence. Lerner [16] (pp. 150-163) gives an excellent discussion of the increasingly complicated, bizarre, and arbitrary hypotheses that have been developed since 1965 to accommodate the more and more evident conflicts of the Big Bang theory with observed data.

Editorial Article in New Scientist Magazine. The May 22, 2004 *New Scientist* (p. 20) has an editorial article criticizing the Big Bang theory, written by Eric Lerner and endorsed by 33 other scientists from 10 countries. The names of scientists endorsing this article are listed in the website ***www.cosmologystatement.org***. The names of other scientist endorsing the article have been added, giving a total of 268 scientists as of March 16, 2005.

This article reports that funding for research on cosmology is tightly controlled by peer-review committees that are dominated by supporters of the Big Bang theory. Dissent against the Big Bang theory is firmly suppressed in conferences. Scientists, "who doubt the big bang, fear that saying so will cost them their funding".

This article explains that the evidence against the Big Bang theory has grown steadily year-by-year, yet is continually dismissed by postulating a steady flow of arbitrary physical assumptions that are unsupported by observational evidence. The "big bang theory can boast of no quantitative predictions that have since been validated".

The wide support of this article by scientist throughout the world demonstrates the growing awareness that there is a severe lack of objectivity in astronomical research.

Chapter 10

The Yilmaz Cosmology Theory

Assumptions of the Yilmaz Cosmology Theory

When Yilmaz presented his theory of gravity in 1958, which was published in the prestigious *Physical Review* [Y1], he applied his theory to the following simple cosmology model:

The universe has a constant average density of matter that extends to infinity and does not change with time.

Yilmaz discovered to his surprise that his gravitational theory predicts an expanding universe. **It predicts that the Hubble expansion of the universe is a natural relativistic effect directly caused by gravity.**

The Yilmaz theory predicts that the Hubble expansion rate of the universe is proportional to the square root of the average density of matter. Appendix C (Section C.6) shows that an average density equivalent to 4.8 hydrogen atoms per cubic meter would produce our nominal Hubble constant (20 km/sec per million light years). As shown in Section C.5, this is highly consistent with the measured density, 2 hydrogen atoms per cubic meter, derived from astronomical data, when one considers the great uncertainty in the measured density.

Astronomical observations show that the average density of luminous matter in the universe is equivalent to 0.006 hydrogen atoms per cubic meter. The total matter in the universe is estimated by observing the rotation of large galaxy clusters. In order for the formation of a galaxy cluster to be stable, there must be about 300 times as much dark matter in the cluster (which we cannot see) as there is luminous matter (which we can see). Multiplying 0.006 by 300 gives about 2 hydrogen atoms per cubic meter for the total matter.

Setting the average density of matter equal to 4.8 hydrogen atoms per cubic meter yields a unique cosmological solution. This chapter presents the plots derived from that solution, which give a remarkable picture of our universe.

In Fig 10-1, the solid curve shows the velocity of a galaxy verses the true distance to the galaxy in billions of light years that is predicted by the Yilmaz cosmology theory. The dashed line shows for comparison the Hubble law, assumed in Big Bang theories, in which galaxy velocity is exactly proportional to distance. The Hubble law assumes that the velocity of a galaxy reaches the speed of light at a distance of 15 billion light years, and exceeds the speed of light at greater distances.

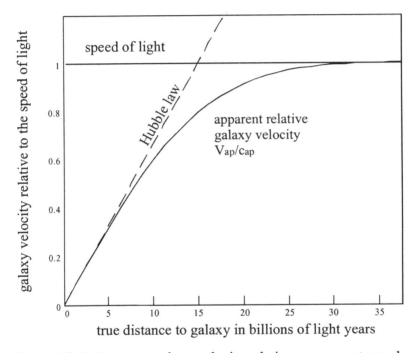

Figure 10-1: Apparent galaxy velocity relative to apparent speed of light, compared with Hubble law.

As shown by the solid curve in Fig. 10-1, the Yilmaz cosmology theory predicts that the velocity of a galaxy never reaches the speed of light. Within 5 billion light years, the galaxy velocity closely follows the Hubble law, but at larger distances it departs drastically from the Hubble law. At very large distances the galaxy velocity gradually approaches the speed of light, but never exactly reaches it.

Compression of the Universe

As we saw in Chapter 8, the Yilmaz theory predicts that a gravitational field causes the speed of light, a clock rate, and a spatial dimension to decrease. The Einstein theory also has these effects, except that these parameters all go to zero at the Schwartzschild limit.

When the Yilmaz theory is applied to cosmology, it predicts that matter in the universe causes the speed of light, a clock rate, and a spatial dimension to decrease as we look out into space. These effects are shown in Fig. 10-2. The solid curve shows how the speed of light decreases with distance from the earth, and the dashed curve shows how a clock rate and a spatial dimension decrease with distance. These parameters never go to zero. However, the speed of light is very small beyond 35 billion light years; a clock rate and a spatial dimension are very small beyond 45 billion light years.

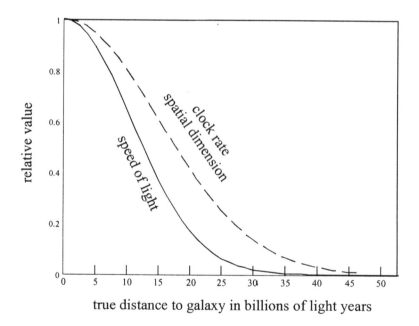

Figure 10-2: Apparent speed of light (solid), clock rate (dashed), and spatial dimension (dashed), versus distance to a galaxy

The horizontal scale of Fig. 10-2 shows the true distance to a galaxy expressed in billions of light years. The Big Bang theory assumes that our observable universe extends only to 15 billion light years; beyond that limit galaxies are theoretically receding faster than the speed of light and so cannot be seen. In contrast, the Yilmaz cosmology model predicts that galaxies can be observed at true distances much greater that 15 billion light-years.

The plot of spatial dimension in Fig 10-2 shows that a dimension appears to contract with distance. Consequently, the apparent distance to a galaxy is less than the true distance. Figure 10-3 gives a plot of the apparent distance to a galaxy versus the true distance. Because of the strong contraction of spatial dimensions at great distances, the maximum apparent distance to a galaxy is finite, even though the model assumes that true galaxy distances extend to infinity. The maximum apparent distance to a galaxy is equal to $\sqrt{[\pi/2]}$ times the radius of the observable universe (15 billion light years) as defined by the Big Bang theory. The maximum apparent galaxy distance of the Yilmaz cosmology theory is $\sqrt{[\pi/2]}$ times 15 billion light years, which is 18.80 billion light years.

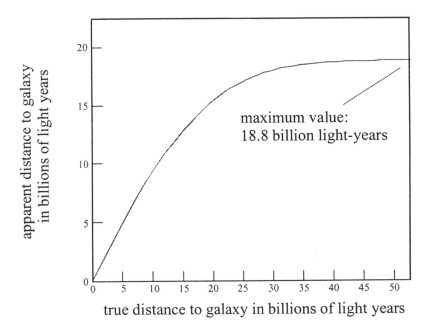

Figure 10-3: Apparent distance to galaxy verses true distance

For a galaxy that is close to the apparent limit of 18.8 billion light years, there is strong compression of the apparent size of the galaxy. Consequently, the apparent density of matter becomes very high as the limit of 18.8 billion light years is approached. This effect is displayed in

Fig. 10-4, which shows how the apparent density of the universe increases with distance.

Figure 10-4 shows how the universe should appear from earth as predicted by the Yilmaz cosmology theory. The earth is the dot at the center. The inner circle is at an apparent distance of 7.5 billion light years, and the periphery is at an apparent distance of 18.8 billion light years. The darker the picture, the greater is the apparent density of the universe. The universe is quite regular out to 7.5 billion light years, where the density has increased by 50 percent. At the outer limit the density is extremely high.

Uniqueness of Yilmaz Theory Predictions

Many different cosmological theories have been derived from the Einstein gravitational field equation. A fundamental problem with the Einstein theory is that its predictions are not unique. When the Einstein theory is applied, multiple, contradictory solutions can be derived from the same physical model. Constraints that are somewhat arbitrary must be included in a general relativity analysis to achieve a solution.

In contrast, the predictions of the Yilmaz theory of gravity are unique. For a given physical model, there can be only one solution. This principle may be difficult to understand by a scientist skilled in applying the Einstein theory. An analysis using the Yilmaz theory does not include the arbitrary adjustable parameters required in an analysis using the Einstein theory.

Since the Yilmaz theory of gravity gives a unique solution to a given physical model, the plots presented in this chapter are not arbitrary. They are not affected in any way by postulates made by the writer, who calculated them. These plots are uniquely constrained by (1) the mathematical validity of the Yilmaz theory of gravity, (2) the very simple cosmological postulate given at the beginning of this chapter, and (3) specifying the average density of matter to achieve a Hubble constant of 20 km/sec per million light years of galaxy distance.

Remember that the Yilmaz theory of gravity applies the principles of the Einstein general theory of relativity, and was derived in a rigorous manner from those principles. The Yilmaz theory is a refinement of the Einstein theory.

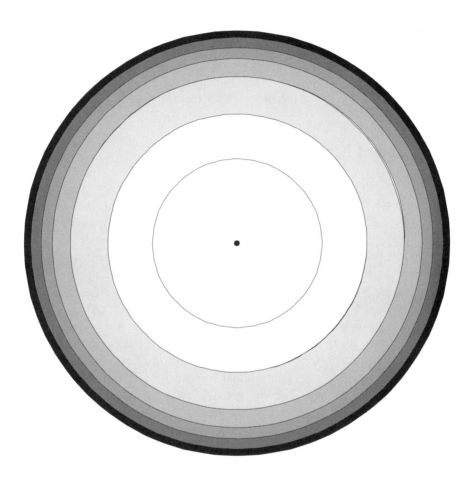

Figure 10-4 Apparent relative mass density of universe seen from earth; boundaries at density values of 1.5, 3, 10, 30, 100, and 1000; minimum and maximum radii at 7.5 and 18.8 billion light years

Since these plots of the Yilmaz cosmology theory were derived by rigorous analysis, they should be taken seriously. They were not affected by arbitrary postulates made in deriving them. This does not prove that these plots give a physically accurate picture of our universe. Nevertheless, there is strong reason to believe that the principles of relativity developed by Einstein are basically correct. These plots were derived rigorously from the relativity principles established by Einstein, even though they are inconsistent with the Einstein gravitational field equation.

Creation of Matter

Figure 10-1 showed that the Yilmaz cosmology theory predicts that the universe must expand. However, the physical assumption on which the cosmology theory is based requires that the average density of matter must stay constant. To satisfy this requirement, matter must be continuously created throughout the universe to compensate for the universe expansion.

This shows that the Yilmaz cosmology theory satisfies the basic principle of the Steady State cosmology concept. Continual creation of matter is a **postulate** of the Hoyle Steady State theory, but is a **prediction** of the Yilmaz cosmology theory. When the Yilmaz gravitational theory is applied to cosmology, it requires that the universe expand, and it requires that matter be continuously created to offset the universe expansion. ***When the Yilmaz gravitational theory is applied to cosmology, it leads inevitably to the Steady State cosmological concept of the universe.***

Constancy of the Universe Size

Figure 10-1 showed that the receding velocity of a galaxy approaches the speed of light at large true distances. However, Fig. 10-2 showed that the speed of light becomes very small at large true distances. Consequently the actual velocity of a galaxy becomes very small, even though the galaxy is traveling at nearly the speed of light.

Figure 10-5 shows the true velocity of a galaxy versus true distance that is predicted by the Yilmaz cosmology theory. The galaxy velocity is expressed in absolute units (kilometers per second). The true galaxy velocity closely follows the Hubble law out to 5 billion light years. It reaches a maximum velocity of 150,000 km/sec at a true distance of 10 billion light years. This maximum velocity is half of the speed of light measured on earth (300,000 km/sec). At very great distances the true velocity of a galaxy becomes very small.

Thus, Fig. 10-5 shows that over very great distances the expansion of the universe approaches zero. This indicates that even though the universe expands about every point, the size of the universe remains constant.

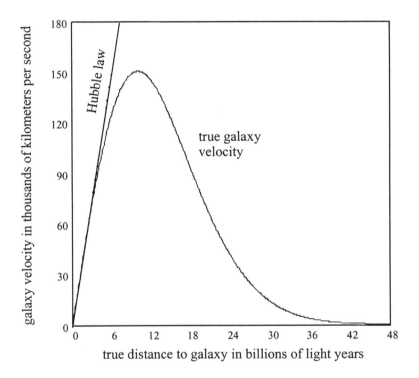

Figure 10-5: True galaxy velocity expressed in thousands of kilometers per second compared with Hubble law.

"How can this be?", you ask. "How can the universe expand everywhere without getting bigger?" The answer is that relativistic effects due to mass in the universe produce strong compression of dimensions at great distances. This compression offsets the expansion of the universe, and so the universe does not become any larger even though it expands everywhere.

This may seem like a strange answer. However, all relativistic effects are strange. We saw in Chapter 6 that when light travels from one observer to another, both observers measure exactly the same speed of light, regardless of the relative velocity between them. That relativistic effect does not seem to make sense, yet is essential for explaining speed of light measurements.

Source of the Continual Creation of Matter

What is the source of the matter that is continuously created throughout the universe? The Yilmaz theory of gravity requires that the sum of matter-plus-energy within the universe must remain constant. Matter can be converted into energy, and vice versa, in accordance with the Einstein formula ($E = Mc^2$), but the sum of matter-plus-energy cannot change. To satisfy this requirement, the diffuse matter continuously created throughout the universe must be derived from energy that is radiated from matter within the universe.

What could be the source of the energy that is creating the diffuse matter. The writer postulates that black dwarf and neutron stars generate gravitational waves, which gradually convert the matter of these stars into radiated energy. The extreme density of black dwarf and neutron stars may accelerate the production of gravitational waves. It is postulated that the radiated energy in these gravitational waves is converted into hydrogen atoms, which form the diffuse matter that is created throughout the universe.

Both the Einstein and Yilmaz theories predict gravitational waves. As explained in *Scientific American*, April 2002 (pp. 62-71), scientists are performing elaborate experiments that are attempting to measure these waves. Hence the postulate that black dwarf and neutron stars continually radiate gravitational waves is consistent with present scientific knowledge.

This gravitational wave postulate yields the following picture of the universe, which explains how our universe can remain eternally young although it is infinitely old. Dying stars produce black dwarf and neutron stars. The matter in these extremely dense stars is gradually converted into energy in the form of gravitational waves, and so the matter in these stars is slowly dissipated.

The energy in the gravitational waves is converted into diffuse matter throughout the universe, which is created at a rate of one hydrogen atom per year within a volume of one cubic kilometer. The hydrogen atoms congregate into huge clouds, which form new stars and galaxies, and the process continues indefinitely.

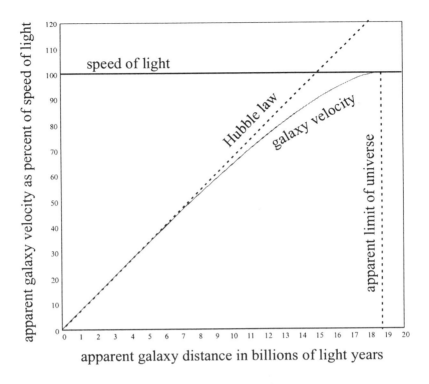

Figure 10-6: Apparent velocity of a galaxy as a percent of the speed of light, versus the apparent distance to the galaxy in billions of light years; also showing the apparent limit of the universe at 18.80 billion light years.

Apparent Variation of Galaxy Velocity with Distance

The plot in Fig. 10-1 on page 149 expresses the apparent expansion of the universe in terms of the true distance to a galaxy. However, that is not what an astronomer sees. An astronomer observes the apparent velocity of a galaxy versus the apparent distance to the galaxy, as shown in Fig. 10-6. The solid curve shows the apparent velocity of a galaxy as a percentage of the speed of light. This is compared with the Hubble law, shown by a dashed line. The Hubble law reaches the speed of light at 15 billion light years. The actual galaxy velocity reaches the speed of light at the *apparent limit of the universe*, which was shown in Fig 10-4 to be 18.80 billion light years.

Astronomers measure the apparent velocity of a galaxy from the redshift of its spectrum. This measurement yields the apparent galaxy velocity as a fraction of the speed of light. The apparent distance of a galaxy billions of light years away is best measured by searching for exploding Type 1a supernovas, which generate a repeatable peak brightness of about 3 billion suns for a couple of weeks. These supernova measurements have been performed by two astronomer teams led by Brian Schmidt and Saul Perlmutter. These studies found that the distant supernovas are further away than the Hubble law predicts from their redshifts. This discovery astounded the astronomers, and led them to conclude that the expansion of the universe must be accelerating.

However, Fig. 10-6 indicates that this astounding finding is just what one would expect from the Yilmaz Steady-State cosmology theory. For a given apparent velocity, the apparent distance of a very distant galaxy is greater than what is measured from the dashed Hubble-law plot.

The Big Bang theory has limitless variable factors that can allow astronomers to reconcile nearly any astronomical measurement with the theory. In contrast, the Yilmaz theory can yield only one answer. If the measurements now being made on supernovas depart appreciably from the solid curve in Fig 10-6, the Yilmaz cosmology theory must be wrong. The Yilmaz theory does not have any adjustable parameters to compensate for discrepancies. *Alternatively, if the measurements on supernovas match the predicted curve in Fig. 10-6, such a match would strongly support the Yilmaz Steady State cosmology theory.*

Cosmic Microwave Background Radiation

Proponents of the Big Bang theory have claimed that *cosmic microwave background radiation* proves the validity of the Big Bang theory. The original Steady State theory proposed by Fred Hoyle did not provide an explanation for this radiation. However, the Yilmaz Steady-State cosmology theory predicts cosmic microwave background radiation more accurately than does the Big Bang theory.

Figure 10-1 shows that at great distances the Yilmaz cosmology theory predicts that the velocity of a galaxy gradually approaches the speed of light, but never quite reaches it. This indicates that the radiation from very distant galaxies should be strongly shifted to low frequencies.

This effect is analyzed in Appendix D. The analysis shows that the light radiated from distant galaxies should produce cosmic microwave background radiation that is equivalent to the radiation from an ideal blackbody at a temperature between 2.1 and 3.4 degrees Kelvin. This prediction is highly consistent with the 2.73 degree Kelvin blackbody temperature actually measured by the COBE satellite.

How Can Gravity Make the Universe Expand?

The Einstein general theory of relativity applies the Riemann-Ricci mathematical theory of curved space, which is based on the geodesic, the shortest distance between two points. The geodesic equation is a fundamental theorem of the Riemann-Ricci mathematical theory. When the geodesic equation is applied to the Yilmaz theory, it predicts the expansion of the universe that was displayed in Fig. 10-1.

Using the geodesic equation, Yilmaz proved in 1958 that his gravitational theory predicts an expanding universe: the Hubble expansion is a natural relativistic effect directly caused by gravity. Nevertheless, even in recent years, Yilmaz found this conclusion to be disturbing. He asked, "How can gravity (which always causes attraction between masses) force the universe to expand?" Let us consider a physical answer to this question that seems to make sense intuitively.

Remember that the expansion of the universe predicted by the Yilmaz theory is a local effect. Over very large distances the universe does not expand. Thus, we need an intuitive explanation of how gravity can make the universe expand locally.

Figure 10-4 shows that the universe appears to be compressed within a sphere with a radius of 18.8 billion light years. The apparent density of matter is extremely high near the periphery of this sphere. This outer shell of high-density matter should exert gravitational force on matter inside the sphere, thereby pulling the matter toward the periphery. In this way, gravitational attraction could cause a local expansion of the universe.

However, this explanation is not consistent with Newton's theory of gravity. Suppose that the universe is modeled as a thin spherical shell, with matter evenly distributed over the shell. If Newton's gravitational theory is applied to a mass element inside the shell, the gravitational forces on the mass element would exactly cancel. There would be no net force attracting the mass element toward the spherical shell.

Nevertheless, the geodesic equation proves that relativistic gravitational effects force the universe to expand locally. The author's intuitive explanation of this process is that gravitational attraction from the high density of apparent mass at the periphery is pulling the universe apart; that gravitational attraction is causing the universe to expand locally.

Chapter 11

Evidence of the Mystery of Creation

What should one believe about the mystery of Creation? To help the reader decide what to believe, the book separates those aspects of Creation that are based on strong scientific evidence from those for which the supporting evidence is weak. Let us first review the concepts of Creation that have solid scientific support.

Concepts of Creation Supported by Strong Evidence

Creation of Life on Earth

The earliest earth rocks have been dated with high reliability to be about 4.3 billion years old. Therefore we have high confidence that our earth has existed as a solid body for at least 4 billion years.

Fossils show a progression of life starting with microscopic cells that appeared about 3.5 billion years ago. After billions of years of microscopic life, the first multi-celled animals appeared about 600 million years ago. There was an explosion of animal life about 550 million years ago at the start of the Cambrian geological period. Beginning with primitive fish about 500 million years ago, we can trace the evolution of vertebrates to advanced fish, to amphibians, to reptiles, and eventually to the modern mammals and birds that dominate our natural world today. This process culminated in the creation of modern humans, which first appeared at least 100 thousand years ago.

This progression of life can be explained by the Darwin theory of evolution. The great consistency of DNA in different species also supports the evolution theory. The writer has not found a coherent explanation for the progression of life over the ages, and the similarity in the cells of different species, that does not accept the principle of evolution.

On the other hand, there are appreciable gaps in our understanding of the creation of life. How did the first life begin? As shown in Appendix A, the DNA code of even the simplest of living cells is highly complicated. How was this DNA coding mechanism created?

The Creation of our Sun and its Solar System

Astronomical studies of stars in various stages of development, and the studies of the nuclear processes that generate the energy of a star, have provided a reliable picture of how our sun was created, how it lives, and how it will eventually die.

There is strong evidence that our sun was created from a huge cloud that was mostly hydrogen gas, but contained some dust particles that made up the solid matter from which our earth and other solid planets were formed. This cloud collapsed because of gravity, and the release of gravitational energy heated the cloud. The temperature and pressure at the center of the cloud became sufficient to ignite nuclear fusion, and our sun began to shine. Nuclear fusion converts hydrogen into helium, and releases an enormous amount of energy. Our sun has been burning hydrogen to form helium for about 5 billion years, and will continue this process for another 5 billion years until the hydrogen fuel is exhausted.

As our sun was created, a solar system of planets and smaller bodies was created in orbits around the sun. The planets that orbit our sun have only 0.13 percent of the mass of the sun, but have 50 times the sun's angular momentum. As the huge cloud of hydrogen gas collapsed to form our sun, the central (sun) portion of the cloud had to lose angular momentum, in order for the cloud to continue to collapse. This was apparently achieved by a process that transferred angular momentum from the central (sun) portion of the cloud to a surrounding disk that eventually formed our planets.

This argument indicates that the formation of a solar system around a star is probably a normal aspect of the creation of a star. Consequently, we should expect that many stars within our Milky Way Galaxy have solar systems with planets physically similar to our earth. However, stars with earth-like planets may be at such enormous distances that we have no means of reaching them.

The planets of our solar system lie close to the same plane. This suggests that the material that formed our planets was originally carried as gas and dust within a thin disk that orbited our sun.

How was angular momentum transferred from the central (sun) portion of the cloud to the surrounding disk? Nobel laureate Hannes

Alfven has proposed that the gas in the disk was ionized to form plasma, with the electrons separated from the atoms. These electrons formed electric currents that produced a gigantic magnetic field. The magnetic field of the disk reacted against the magnetic field of the central (sun) portion of the cloud, and the resultant torque transferred angular momentum from the sun to the surrounding disk.

The astronomical community has rejected the research of Hannes Alfven and other plasma physicists concerning the development of our solar system. Astronomers have ignored the Alfven theory, but do not have an alternate explanation for the process that allowed the sun to lose excess angular momentum as its hydrogen cloud collapsed.

Thus, we have a general picture for explaining how our sun and our earth were created, but many gaps in the story remain to be clarified.

The Creation of the Stars

Astronomical evidence shows that essentially all stars are probably created in the same way. The size of the initial cloud of hydrogen gas determines the mass of the resultant star. Stars vary in mass from about 1/10 to 100 times the sun's mass. The more massive a star, the brighter it shines, and the faster it burns up its hydrogen fuel. A star with twice the sun's mass has about 1/8 of the sun's lifetime, and a star with half of the sun's mass has about 8 times the sun's lifetime.

Stars with more than 8 times the mass of our sun use up their hydrogen fuel in millions of years (rather than in billions of years like our sun) and end their lives in gigantic supernova explosions. These massive stars produce nuclear reactions that convert hydrogen into heavy elements, and the supernova explosions scatter the heavy elements around the galaxy. The heavy elements form dust particles that are picked up in the hydrogen clouds that create new stars. These dust particles are the solid matter from which planets like earth are created.

Thus, the total process of stellar creation apparently began with huge clouds containing only the simplest element, hydrogen. A hydrogen atom consists of a single electron orbiting a single proton. All of the heavier atoms of our universe were apparently created from simple hydrogen atoms primarily by supernova explosions. Starting with clouds of pure hydrogen gas, we can explain with reasonable confidence the creation of all of the elements and nearly all of the stars of our universe. (There are some stars, like quasars, that we do not understand, and therefore we cannot explain how they may have been created.)

This leaves us with the fundamental mystery of Creation. How were the clouds of hydrogen gas created? In our search for answers to this mystery, we leave the domain of solid scientific evidence and enter the domain of speculation.

How Were the Clouds of Hydrogen Created?

The Basic Theories of Universe Creation

There are two basic concepts for explaining the creation of the hydrogen gas that produced the stars of our universe. These are:

(1) *The Big Bang theory,* which postulates that the hydrogen gas was created in a major creation event occurring in the distant past;

(2) *The Steady State theory,* which postulates that hydrogen gas is continually created throughout the universe as diffuse matter, probably one hydrogen atom at a time.

In order for these two concepts to be consistent with the Hubble expansion of the universe (a Hubble constant of 20 km/sec per million light years), the following constraints are added:

(1) *The Big Bang theory:* The creation event occurred as a Big Bang explosion about 15 billion years ago;

(2) *The Steady State theory:* The diffuse matter is created at a rate of one hydrogen atom per year within a volume of one cubic kilometer.

Theory that the Universe Expansion is Apparent

To explain the Hubble expansion of the universe, we considered in Chapter 5 (p. 90) a third cosmological theory, which postulates that the universe expansion is *Apparent*. This theory assumes that the universe is not actually expanding, and that an effect other than velocity is causing a redshift proportional to galaxy distance, which makes the universe appear to expand.

The theory that the Hubble expansion is an *Apparent* effect has the serious problem that no mechanism has been proposed that is quantitatively consistent with the observed Hubble redshift. On the other hand, a more fundamental weakness of this theory is that it does not explain the creation of matter. Our universe is aging, and new stars are being created from clouds of hydrogen gas. What produced the hydrogen from which stars have been and are being created?

Was the hydrogen created in the distant past by a major creation event? Or, is the hydrogen being created continuously throughout the universe as diffuse matter? The theory that the universe expansion is *Apparent* does not attempt to explain the creation of matter. Therefore, even if one endorses this theory that the universe is not expanding, one is still left with the same choices for explaining the creation of matter: (1) the Big Bang theory, and (2) the Steady State theory.

Our Two Basic Theories for Creation of Matter

Thus, we have two general explanations for the creation of the hydrogen gas that seems to be the source of all of the matter of the universe. These are the Big Bang theory and the Steady State theory. Let us examine these cosmological concepts in detail.

The Big Bang Theory

The Gamow Big Bang Theory

The expansion of the universe first observed by Hubble shows that galaxies are flying apart. The obvious interpretation of this is that the whole universe has emerged from a gigantic Big Bang explosion that occurred billions of years ago.

George Gamow recognized that the greatest possible density of matter exists in the neutron star, which has the density of nuclear matter, one billion tons per teaspoon. Consequently, Gamow postulated that our universe began as a single body with the density of nuclear matter, which exploded with a Big Bang.

Modern astronomical measurements yield a Hubble constant of about 20 km/sec per million light years of galaxy distance. Assuming that the universe is expanding at a uniform rate, a galaxy 15 billion light years away would recede at the speed of light, and so cannot be seen. Hence our observable universe is considered to be a sphere with a radius of 15 billion light years.

At that time, Gamow did not know how far the universe extended. However, modern astronomical measurements show that the universe extends for many billions of light years, presumably to the limit of the observable universe.

If we assume that all of the matter within our observable universe was originally compressed within a single sphere with the density of nuclear matter at the instant of the Big Bang, the initial universe would have just about fit within the orbit of the planet Mars. This is the initial size of the observable universe for the Gamow Big Bang theory.

Cosmic Microwave Background Radiation

Gamow calculated that the universe was opaque until 300 thousand years after the Big Bang, and massive optical radiation was emitted at that time. As the universe expanded, the optical radiation would have decreased in frequency, and should be present today at microwave frequencies. This optical radiation from the early universe should be observable today as *cosmic microwave background radiation,* which emanates from all directions and has the spectrum of a blackbody. Gamow made several estimates of the blackbody temperature of this radiation, which varied from 5 to 20 degrees Kelvin.

In 1965, cosmic microwave background radiation was discovered by Penzias and Wilson in a sensitive microwave antenna at Bell Laboratories. The estimated blackbody temperature of the radiation was 3.5 degrees Kelvin. In 1989, the COBE satellite accurately measured this cosmic radiation. The radiation precisely follows the spectrum of a blackbody at 2.73 degrees Kelvin, and arrives with extreme uniformity from all directions. The intensity of the radiation varies with direction by only a few parts in 100,000.

Supporters of the Big Bang theory proclaimed that cosmic microwave background radiation is proof that the Big Bang theory must be correct. The competing Steady State theory proposed by Hoyle could not explain this radiation, and so interest in the Steady State theory rapidly faded.

Nevertheless, as shown in Appendix D, the Yilmaz Steady-State theory described in Chapter 10 predicts cosmic microwave background radiation, and gives a more accurate value for the blackbody temperature of this radiation than the best estimate derived from the Big Bang theory.

The Modern (Singularity) Big Bang Theory

In the mid 1960's, powerful computers became widely available, and many scientists began to use them to study the complicated Einstein equations. About the only area where these studies could be applied was cosmology, and so hundreds of scientists were attracted to theoretical cosmological research, which almost always led to the Big Bang theory.

Big Bang cosmology studies using the Einstein equations invariably predicted that the universe began at the instant of the Big Bang as a singularity having nearly infinite density of matter. In 1994, James Peebles (dubbed by *Scientific American* to be "the father of modern cosmology") concluded that our present observable universe (30 billion light-years in diameter) was initially "smaller than a dime".

Practically all astronomers today accept this constraint by Peebles, but most adopt the "inflation" concept proposed by Alan Guth, which concluded that the initial universe was microscopic in size. Dickinson [28] (p. 118) graphically portrays the initial "inflation" part of the Big Bang explosion, in which our present observable universe expanded from "one trillionth of the size of a proton to the size of a baseball". (A proton is 2 trillionths of a millimeter in diameter.)

Along with the Big Bang singularity, astronomers also claim that similar singularities exist today within black holes. They have observed massive dense stars, which, according to the observations, could be massive neutron stars. However, as reported by Dickinson [28] (p. 89), astronomers insist that these massive dense bodies cannot be neutron stars. Since the Einstein equations predict that a massive neutron star must collapse into a black hole, astronomers conclude that these massive bodies must be black holes. On the other hand, a black hole severely violates physical evidence, because the star within a black hole must be squeezed into a singularity having an infinite density of matter.

Einstein's Rejection of the Singularity

The Einstein theory is the sole basis used by astronomers to justify the Big Bang and black hole singularities. Nevertheless, Einstein absolutely rejected the singularity concept throughout his lifetime.

As was shown in Chapter 7, Einstein rejected the black hole singularity in 1939. Einstein's argument that rejected the black hole singularity was not challenged during his lifetime. However, about a decade after his death, computer studies of the Einstein equations were used to prove that Einstein's 1939 argument was wrong. Hence, scientists accepted the claim that the Einstein equations do indeed predict a black hole singularity.

This point is illustrated in the following quotation (given earlier in Chapter 5, p. 96), which was presented in a book by Filkin [17] (p. 104):

11. Evidence of the Mystery of Creation   167

"Stephen [Hawking] and Roger Penrose published a paper in 1970 which proved that, if Einstein's mathematics were correct, a singularity had to result from a black hole, and had to exist at the start of the universe. - - - The paper argued that if relativity as explained by Einstein is correct — and all of the evidence from observation seems to keep confirming it — then the universe must have started with a big bang explosion out of a singularity. The equations do not allow an alternative."

Thus, scientists concluded from computer studies that the Einstein equations definitely predict black hole and Big Bang singularities.

In 1945, Einstein recognized that his theory implied a singularity at the instant of the Big Bang. Let us reexamine the following quotation by Einstein, which was given in Chapter 7 (p. 128):

"Theoretical doubts [concerning the creation of the universe] are based on the fact that [at the] beginning of the expansion, the metric becomes singular and the density becomes infinite. . . In reality, space will probably be of a uniform character, and the present [relativity] theory will be valid only as a limiting case. . . **One may not therefore assume the validity of the equations for very high density of field and of matter, and one may not conclude that the 'beginning of the expansion' must mean a singularity in the mathematical sense.** *All we have to realize is that the equations may not be continued over such regions."*

Einstein stated in this quotation that his equations are only approximately correct under conditions of very high density of field and of matter, and so cannot be used to predict a physical singularity at the instant of the Big Bang. Applying Einstein's reasoning, we should also conclude that (according to Einstein) the Einstein theory cannot be used to predict a black hole singularity.

The statement by Hawking and Penrose given above said that, "all of the evidence from observations seems to keep confirming [Einstein's relativity theory]". However, as was shown in Fig. 6-6 of Chapter 6, the gravitational effects of the Einstein theory that can be measured within our solar system are extremely small. Validation of these tiny effects does not prove that the Einstein equations apply accurately in the region near the Schwartzschild singularity, where the mass/radius ratio is 240,000 times the maximum value occurring within our solar system.

Einstein maintained that his equations would only hold

approximately under conditions of extreme density of matter.

The Big Bang Age Dilemma

A serious problem with all versions of the Big Bang theory is that it does not allow sufficient time to explain the development of the universe. As was shown in Chapter 5 (p. 93), there is strong evidence that at least some stars within our Milky Way galaxy are at least 13.4 billion years old. Nevertheless it is generally concluded that the Big Bang occurred 13.7 billion years ago, no more than 300 million years before these stars were created.

The discussion earlier in this chapter showed that cosmic microwave background radiation implies an extremely uniform universe 300,000 years after the Big Bang. There is simply not enough time for our present lumpy universe to have been created in the short time since the Big Bang. (See Chapter 5, p. 93.)

The Steady State Theory

The Hoyle Steady State Theory

In 1948, Fred Hoyle explained the Hubble expansion of the universe by postulating that diffuse matter is being continuously created throughout the universe at a rate sufficient to compensate for the universe expansion, so that the average density of matter remains constant. With this assumption, Hoyle postulated that the universe is infinitely old, and has always appeared to be more or less like we see it today.

Hoyle did not propose a clear mechanism that might be creating the diffuse matter throughout the universe. He postulated that when a galaxy exits our observable universe, there is a cosmological process that converts the mass that is leaving our universe into new diffuse matter within the universe.

When cosmic microwave background radiation was discovered in 1965, Hoyle could not explain it with his Steady State theory. Consequently, interest in the Hoyle theory rapidly faded.

The Yilmaz Steady State Theory

The Einstein general theory of relativity does not yield a unique solution for a particular physical model. Consequently many different cosmology theories have been derived from the Einstein theory, including the Hoyle Steady State theory and the many variations of the Big Bang theory.

The Yilmaz gravitational theory is a refinement of the Einstein theory, which applies the principles established by Einstein in developing his theory. However, the Yilmaz theory does not accept the Einstein gravitational field equation, which Einstein obtained in an intuitive manner. The Yilmaz theory is more strongly constrained than the Einstein theory, and can yield only a single solution for a particular physical model.

When the Yilmaz gravitational theory is applied to cosmology, it uniquely yields a Steady State universe solution. *The expansion of the universe is predicted by the Yilmaz theory to be a natural relativistic effect.* By assuming that the average density of matter of the universe is equivalent to 4.8 hydrogen atoms per cubic meter, the Yilmaz theory predicts that the universe must expand at our measured Hubble rate, 20 km/sec per million light years of galaxy distance.

As shown in Appendix D, the Yilmaz theory predicts that the optical radiation from very distant galaxies should produce strong cosmic radiation at microwave frequencies, which is equivalent to the radiation from a blackbody at a temperature between 2.1 and 3.4 degrees Kelvin. This result is closer to the measured cosmic microwave background radiation obtained by the COBE satellite (2.73 degrees Kelvin) than the best estimate derived from the Big Bang theory.

Chapter 10 described the Yilmaz Steady State cosmology theory, showing that it gives remarkably consistent answers for the creation of matter in our universe.

The Yilmaz gravitational theory does not allow a singularity, either a Big Bang singularity or a black hole singularity. *All of the astronomical evidence justifying black holes can be readily interpreted as evidence for massive neutron stars.*

We Do Not Know the Answer

For many years astronomers have confidently claimed that they know that all of the matter of our universe was created billions of years ago in an enormous Big Bang explosion. However, when we investigate this claim, we find that the Big Bang theory is highly speculative. The weakness of the theory is much more apparent in its modern versions, which are based on the singularity principle. Modern Big Bang theories all predict that the universe began as a singularity of infinitesimal size, a concept that radically violates physical evidence.

Along with the Big Bang singularity, astronomers endorse the black hole singularity, which requires that all of the matter inside a black hole must be squeezed into a state of infinite density.

Despite the overwhelming support by the astronomical community of these bizarre predictions, there is no justification for treating them as scientific fact. In the field of astronomy today, consensus is being used as a substitute for scientific evidence. Astronomers apparently think that they need an answer for everything. The scientific approach should be to admit that *we do not know the answer*.

This book has presented the Yilmaz Steady State cosmology theory to explain the creation of matter in our universe. Whether or not one accepts the Yilmaz cosmology theory, it demonstrates that there is at least one viable alternative to the unbelievable Big Bang singularity. When astronomers refer to the Big Bang event as an established fact, a common occurrence today, they are severely violating scientific principles. They should treat the Big Bang concept as a theory, and welcome investigations that contradict that theory.

What Should the Reader Believe about Creation?

The issue of Creation is one of the greatest mysteries of life. The information presented in this book should help the reader determine what to believe about this age-old mystery.

For some aspects of Creation, the scientific evidence is strong. For others, the evidence is weak and contradictory. Readers should decide for themselves what claims to accept and what to reject.

In medieval times the people looked to the Church to find answers to this great question, but that philosophy is scorned today. However, one wonders if the proclamations of the Church were not closer to scientific reality than many claims we hear today from scientists. The concept that the stars rotate around the earth each day seems to make more sense than the claim that our whole universe began as a singularity "one trillionth of the size of a proton".

How can the reader find the truth? You should use your own judgment. A layperson can be a better judge of scientific truth than a scientist who is controlled by economic interests to follow a bandwagon of accepted beliefs.

APPENDICES

Appendix A

The DNA Genetic Code

In 1953, James D. Watson and Francis H. Crick discovered the structure of DNA, which specifies the inheritance characteristics of all living organisms. Fifty years later, on April 14, 2003, scientists completed the process of mapping the 3 billion components in the DNA genetic code for a human.

The genetic code carried by the strands of DNA is established by the four nucleotides called A, C, G, and T (adenine, cytosine, guanine, and thymine). The sequences of these four basic components establishes the physical characteristics of all living organisms, including humans.

How does an organism use this DNA information to construct a living cell? A key step in this process is the coding procedure for translating the A, C, G, T nucleotides into the 20 amino acids that control the development of the cell. These amino acids are:

(1) Phenylalanine	(6) Serine	(11) Histidine	(16) Glutamate
(2) Leucine	(7) Proline	(12) Glutamine	(17) Cysteine
(3) Isoleucine	(8) Threonine	(13) Asparagine	(18) Tryptophan
(4) Methionine	(9) Alanine	(14) Lysine	(19) Arginine
(5) Valine	(10) Tyrosine	(15) Aspartate	(20) Glycine

The DNA strands do not directly produce the amino acids. There is an intermediate genetic material, called RNA, which copies the code in the DNA strands, and the RNA material controls the production of the 20 amino acids. The RNA copies the nucleotides in the DNA strands, but replaces each T (thymine) component with U (uracil). The RNA for humans consists of 3 billion nucleotides called A, C, G, and U.

172   Beliefs and Facts of Creation

The A, C, G, U components of RNA are arranged in groups of 3 nucleotides, where each group of 3 nucleotides specifies one of the above 20 amino acids. There are 4x4x4, or 64, possible combinations of the 4 nucleotides within each 3-nucleotide group. Table A-1 shows the coding relationship used in human cells for relating RNA nucleotides to the 20 amino acids. This table holds for nearly all organisms, but some organisms have somewhat different coding relationships. Three of the sequences produce a "stop" command, which represents a break in the amino acid sequence. [64]

Table A-1: *Coding relationships used by humans and most living organisms to relate RNA elements to the amino acids*

UUU (1)	UCU (6)	UAU (10)	UGU (17)
UUC (1)	UCC (6)	UAC (10)	UGC (17)
UUA (2)	UCA (6)	UAA stop	UGA stop
UUG (2)	UCG (6)	UAG stop	UGG (18)
CUU (2)	CCU (7)	CAU (11)	CGU (19)
CUC (2)	CCC (7)	CAC (11)	CGC (19)
CUA (2)	CCA (7)	CAA (12)	CGA (19)
CUG (2)	CCG (7)	CAG (12)	CGG (19)
AUU (3)	ACU (8)	AAU (13)	AGU (6)
AUC (3)	ACC (8)	AAC (13)	AGC (6)
AUA (3)	ACA (8)	AAA (14)	AGA (19)
AUG (4)	ACG (8)	AAG (14)	AGG (19)
GUU (5)	GCU (9)	GAU (15)	GGU (20)
GUC (5)	GCC (9)	GAC (15)	GGC (20)
GUA (5)	GCA (9)	GAA (16)	GGA (20)
GUG (5)	GCG (9)	GAG (16)	GGG (20)

Although most organisms follow the coding arrangement shown in Table A-1, scientists have discovered 16 variations of this coding table. For example, most organisms translate the RNA sequence CUG to produce the amino acid Leucine (2), but many species of the fungus Candida interpret CUG to produce Serine (6).

The amino acids, produced by the RNA code, control the development of each cell by generating different biological molecules, and stopping the generation when enough have been produced.

A virus cell contains RNA but does not have DNA. Consequently, a virus must invade another more complicated cell having DNA in order to reproduce. The virus uses the reproductive mechanism of the complicated cell to reproduce itself, and in the process destroys the other cell.

All truly living organisms have RNA, and, except for viruses, also have DNA. However, there is a strange semi-organism, called a *prion*, which has protein molecules that can reproduce themselves without RNA or DNA. *Mad Cow Disease* is caused by a *prion*. A prion is not truly alive, but it has many of the characteristics of a living organism. [65]

There is remarkable consistency between the DNA and RNA processes of all living organisms. This appears to show that all living organisms are biologically related. Even if one questions this broad conclusion, it seems impossible to deny that humans are biologically related to the chimpanzee. Of the 3 billion nucleotides of the chimpanzee DNA sequence, 95 percent match that of the human. [63]

Appendix B

The Nature of Matter

Our knowledge of the atom evolved from extensive physical research. The following summary shows that this knowledge is drastically inconsistent with the singularity predictions of modern Big Bang cosmologists. This material is based primarily on the excellent explanation of the atom given by Asimov. [31]

What is a Molecule?

Matter is made up of tiny elements called molecules, which in turn are made up of atoms. The character of a molecule is best displayed when matter is in gaseous form. Consider a cubic centimeter of gas at atmospheric pressure and room temperature (20 °C). This volume contains 25 billion times one billion molecules. The number of molecules in this volume is approximately the same for all gases.

The number of molecules is proportional to PV/T, where P is the pressure, V is the volume, and T is the temperature in degrees Kelvin. Kelvin temperature is the number of degrees Celsius measured above absolute zero temperature, which is -273 °C. At absolute zero temperature the random motion of molecules is zero, and so a gas at absolute zero temperature would not exert pressure if it remained a gas.

Elementary Particles that Form an Atom

The atom consists of *electrons*, *protons*, and *neutrons*. All three have been observed as separate particles. Unlike the electron and the proton, the neutron is unstable as a separate particle. Outside the atom, a neutron decays, with a half-life of 12 minutes, into a proton plus an electron. This means that over a period of 12 minutes half of the neutrons decay. When a neutron decays, the very elusive *neutrino* (with an extremely small mass traveling at close to the speed of light) is emitted along with a proton and electron. The protons and neutrons of an atom are contained within a very compact nucleus. A cloud of electrons orbits around the nucleus.

The number of protons in an atom is called its atomic number. The atomic number of an atom can be measured by X-rays. When an electron beam strikes matter, each atom radiates an X-ray with a frequency proportional to its atomic number. This frequency can be measured by directing the X-ray beam against a crystal. The regular pattern of atoms in the crystal diffracts the X-ray beam, and from this diffraction pattern one can determine the X-ray wavelength. From this wavelength one can find the number of protons within the atom. In a normal (non-ionized) atom, the number of electrons is equal to the number of protons.

By using a compound of an atom in gaseous form, one can measure the mass of each molecule, and from this one can determine the mass of each atom (the atomic weight). The atomic weight indicates the total number of particles (protons plus neutrons) in the nucleus. Comparing this total number with the atomic number (the number of protons) gives the number of neutrons.

In this manner, physicists have measured the number of protons, electrons, and neutrons in each atom.

Chemical Reactions among Atoms

The electrons that circle the nucleus of an atom are contained in separate groups of electron orbits called "shells". Heavy atoms have several shells. The number of electrons in the outer shell determines the manner in which the atom behaves chemically.

An atom reaches its most stable chemical state if its outer shell is either full or empty. For hydrogen and helium atoms, which have only one electron shell, the outer shell is full with 2 electrons. For all other atoms, the outer electron shell is full with 8 electrons.

Consider the oxygen atom, which normally has 8 electrons, with 6 electrons in its outer shell. When hydrogen is burned in oxygen, water is formed with the chemical formula H_2O. This indicates that 2 atoms of hydrogen (H) are combined with one atom of oxygen (O). Each hydrogen atom donates one electron to the oxygen atom, thereby increasing the number of electrons in the outer oxygen shell from 6 to 8. The outer oxygen shell is full (with 8 electrons) and the single hydrogen shell is empty in both hydrogen atoms.

In this manner, the atomic model can explain countless chemical reactions that form chemical compounds among atoms.

A normal atom has an equal number of protons and electrons and so has no electrical charge. When an atom gains or loses an electron, it becomes an *ion*, which has an electrical charge. If the atom gains an electron, it becomes a negative ion, and if it loses an electron it becomes a positive ion. When an atom gains or loses an electron, we say that the atom is *ionized*.

The number of protons in an atom determines the chemical properties of the atom. An element can have different forms with the same number of protons (and the same chemical properties) but with different numbers of neutrons. These different forms of an element are called *isotopes*. The main isotope of hydrogen has one proton in the nucleus. *Deuterium* is an isotope of hydrogen with one proton and one neutron. When deuterium is burned in oxygen, it forms "heavy water", which is about 10 percent heavier than normal water.

It is very difficult to separate isotopes of an element, because their chemical reactions are the same. To produce an atomic nuclear bomb with uranium, two uranium isotopes of nearly the same atomic weight must be separated. The uranium is converted into gaseous form as the compound *uranium hexaflouride*. This gas is fed into a centrifuge, which increases the concentration of the heavier isotope at its outer port, and the concentration of the lighter isotope at its inner port. Many stages of this process are required before adequate isotope separation is achieved. Isotope separation is extremely complicated, and is the most difficult part of constructing an atomic nuclear bomb.

Radiation Spectrum of an Atom

When an atom absorbs energy from electromagnetic radiation, an electron moves from a lower to a higher orbit. Later the electron may drop to a lower orbit, and the energy loss is radiated in the form of an electromagnetic wave.

Energy can be absorbed or radiated by an atom only in discrete units called quanta. Each quantum energy change corresponds to a particular wavelength for the electromagnetic radiation that is absorbed or emitted. Consequently, a given atom absorbs or emits radiation only at particular wavelengths.

Absorption and radiation spectra for an atom have been related to the orbits of electrons in the electron cloud that circles the nucleus. As physical studies increased in complexity, it became more meaningful to consider that electrons have particular energy states rather than particular orbits. Nevertheless, the physical model of an electron cloud orbiting a compact nucleus is still supported by the physical evidence.

Experiments with Radioactivity

After radioactivity was discovered in the 1890's, extensive physical experiments associated with radioactive decay were performed. It was discovered that when radioactive atoms decay they emit three different types of rays, which were arbitrarily called *alpha, beta, and gamma rays*.

Gamma rays are electromagnetic waves having wavelengths shorter than X-rays, whereas alpha and beta rays are particles. The *beta ray* is an electron. The *alpha ray* is a helium ion, consisting of 2 protons and 2 neutrons. When alpha ray particles gather electrons, they form helium.

The basic structure of the atom was determined by Ernest Rutherford {1871-1937) in experiments performed between 1906 and 1911. Rutherford bombarded gold leaf with a beam of alpha particles emitted by radioactive atoms. A photographic plate was placed behind the gold leaf to detect the alpha particles. Nearly all of the alpha particles passed through the gold leaf without any effect, even though the gold leaf had the thickness of 20,000 gold atoms (0.005 millimeter). However, a small fraction of the alpha particles were deflected as they passed through the gold leaf.

Rutherford concluded that the path of an alpha particle is not affected when the particle collides with an electron, because an alpha particle has about 7000 times the mass of an electron. The path is only altered when the alpha particle hits a proton or neutron. The gold atom has 79 protons and 118 neutrons. Rutherford concluded that the 197 protons-plus-neutrons are packed together into a very compact nucleus. By determining the fraction of the alpha particles that were deflected after passing through 20,000 atoms, he calculated the size of the nucleus. The diameter of a gold atom is about 40,000 times greater than that of its nucleus.

Based on Rutherford's experiment and many later experiments, it has been found that the density of matter in the nucleus is about 200 million metric tons per cubic centimeter, or about one billion tons per teaspoon. One metric ton is 1000 kilograms and is about 1.1 English tons.

When a radioactive atom emits an alpha particle, it forms a new atom having 2 fewer protons and 2 fewer neutrons. Hence the new atom is an element having an atomic number reduced by 2 and an atomic mass number reduced by 4. When a radioactive atom emits a beta particle (an electron), a neutron is converted into a proton, and so the atomic number of the new atom is increased by one, but the atomic mass number is not changed.

For example, consider the decay of uranium(92)-238 to produce radium(88)-226, and finally lead(82)-206. The value in parentheses is the atomic number (protons), and the value after the dash is the atomic mass number (protons plus neutrons). Uranium(92)-238 emits an alpha particle to form thorium(90)-234, with the atomic number reduced by 2 and the mass number reduced by 4. Two beta particles are emitted in succession to raise the atomic number back to 92 and produce uranium(92)-234. An alpha particle is released to form thorium(90)-230, and this releases an alpha particle to form radium(88)-226. The radium(88)-226 eventually decays in 9 steps, emitting 5 alpha particles and 4 beta particles, to form the stable isotope lead(82)-206.

Uranium-238 decays with a half-life of 4.5 billion years, and so emits very weak radiation. Radium-226 decays with the much faster half-life of 1620 years. Since the earth was formed 4.6 billion years ago, there is still appreciable uranium-238 in the earth. Left by itself, the radium-226 (with its 1620-year half life) would disappear, but the radium-226 is continually being replenished by the decay of uranium-238. Since the half-life of uranium-238 is 2.8 million times the half life of radium-226, the concentration of uranium-238 in the ground is 2.8 million times the concentration of radium-226.

Particle Accelerators

Many nuclear experiments were performed using alpha particles emitted by radioactive atoms to bombard other atoms. One of these was the classic experiment by Rutherford described above, in which he bombarded gold leaf with alpha particles. This approach was extended by using particles accelerated in electrical fields. The usual particle for these atom-smashing experiments is the ionized hydrogen atom, which is a proton. During the 1930's, more and more energetic particle accelerators were built, which often used a circulating design called a "cyclotron".

A neutron is a convenient particle for atom smashing, because it has a greater probability of hitting the nucleus. A proton is repelled by the positive charge of the nucleus. However, it is difficult to accelerate neutrons, because they have no electrical charge. One method is to accelerate an ion of deuterium (heavy hydrogen), which consists of a proton and a neutron that are weakly coupled. As this particle approaches the nucleus of an atom, the proton is repelled and is separated from the neutron, which smashes into the nucleus.

Nuclear Fission

The Italian nuclear physicist, Enrico Fermi, (who emigrated to America) bombarded uranium with neutrons and had a result that he could not explain. In December 1938, the German physicist Otto Hahn repeated this experiment in Berlin, and discovered that the uranium(92) atom was being "split" almost in two to form a barium(56) atom and a krypton(36) atom. The process also released 3 or 4 neutrons.

This was a revolutionary discovery; earlier nuclear experiments had only made small changes of atomic number. This splitting of the atom was called "fission" because it was named after the biological "fission" process by which bacteria split in two when reproducing.

It was immediately recognized throughout the world that this nuclear fission process can release a huge amount of energy, and so might be used in an atomic nuclear power plant and in an atomic nuclear bomb. However, the nuclear bomb seemed dubious. In order to achieve a nuclear bomb, sufficient uranium was needed to achieve the "critical mass" required for a "chain reaction". Early estimates indicated that several tons of uranium were needed.

The neutrons emitted in nuclear fission are moving rapidly. Fast neutrons are needed in a nuclear bomb but not in a nuclear power plant. If a neutron is slowed, its probability of hitting an atom nucleus is greatly increased, because (1) it stays near the nucleus for a longer period, and (2) being slower, its apparent size is much larger. The neutrons in this fission process can be slowed by a "moderator", the most effective being heavy water, made from the hydrogen isotope deuterium. Carbon is also commonly used as a moderator.

In February 1939, Niels Bohr concluded that this fission reaction probably involves the isotope uranium-235. Only 0.7 percent of the atoms in uranium are uranium-235, and it is extremely difficult to separate uranium-235 from the common uranium-238 isotope. This finding led nearly all nuclear physicists to conclude that it was impractical to apply nuclear fission in a bomb.

However, in late 1939 German-Jewish immigrant physicists Rudolph Peirels and Otto Frisch in England discovered that the amount of uranium-235 needed for an atomic bomb is of the order of pounds, not tons. This led to an extensive study, which established by June 1941 the practical principles for developing a nuclear fission bomb. Pilot-plant work commenced on a process for separating the uranium-235 isotope.

This information was transferred from Britain to the United States, and probably advanced the Manhattan development effort by at least a year. The Manhattan nuclear bomb project started in June 1942. The history of early work in developing the bomb is explained by Clark. [32]

Plutonium (atomic number 94) can also split in a chain reaction to achieve a nuclear fission bomb. Plutonium is produced in a nuclear power reactor when neutrons released by uranium-235 collide with atoms of uranium-238 (atomic number 92).. Although plutonium is created in very small amounts, it can be readily separated chemically from uranium-238, because it is a different element. It is extremely difficult to separate the uranium-235 isotope from uranium-238.

The nuclear fission atomic bomb dropped in August 1945 on Hiroshima, Japan used uranium-235, often called *enriched uranium*. Plutonium was used in the nuclear fission bomb dropped a few days later on Nagasaki. The plutonium bomb had not been tested, and so the physicists were not sure it would work.

Nuclear Fusion

Stars derive their energy from nuclear fusion, in which light atoms are fused together to form heavier atoms. The major nuclear fusion process involves the fusion of four hydrogen atoms to form one helium atom. This process releases about 180,000 kilowatt-hours of energy for every gram of hydrogen that is converted into helium. Nuclear fusion operates in the center of the sun at a temperature of 15 million °C and at very high pressure. To achieve nuclear fusion, the temperature and pressure must be sufficient to force the atomic nuclei close together.

After building the nuclear fission bomb, the United States developed the hydrogen bomb, in which a nuclear fission bomb ignites nuclear fusion by applying enormous temperature and pressure to hydrogen. A hydrogen bomb is much more powerful than a nuclear fission bomb, because much more energy is released per gram of material from nuclear fusion than from nuclear fission. The first hydrogen bomb was exploded on November 1952 in the Eniwetok Atoll of the Marshall Islands.

For a half century, extensive research has been performed throughout the world to achieve controlled nuclear fusion in a power plant. This process would not emit harmful radiation, and a vast supply of fuel is available. It would use heavy hydrogen (deuterium), which is found in sea water.. Since the process cannot achieve the enormous pressure at the center of the sun, a temperature of 100 million °C is needed, rather than the 15 million °C occurring in the sun. Unfortunately the goal of controlled nuclear fusion remains beyond our reach.

Energy Released by Nuclear Fusion and Fission

Table B-1 calculates the energy released by various nuclear fusion and nuclear fission processes. (The data are obtained from Ref. [33], Table 7b-3.) The *Atomic Number* (At. No.) is the number of protons in the atom. The *Mass Number* (Mass No.) is the number of protons plus neutrons in the atom for the particular isotope. The *Atomic Weight* is the weight of the particular atom isotope.

The atomic weight in column (4) is divided by the mass number in column (3) to obtain the value in column (5), which gives the atomic weight per nuclear particle (proton or neutron). The differences of the values in column (5) for subsequent elements are shown in column (6), multiplied by 100 to express them in percent.

Table B-1: Atomic Weights of Elements Involved in Nuclear Fusion and Fission

At. No.	Element	Mass No.	Atomic Weight	At. Wt. / Mass No.	Change	KWhr/gm
1	Hydrogen (H)	1	1.007807	1.007807		
					0.710 %	177,500
2	Helium (He)	4	4.0026	1.00065		
					0.065 %	16,250
6	Carbon (C)	12	12.0000	1.00000		
					0.118 %	29,500
26	Iron (Fe)	56	55.9342	0.99882		
36	Krypton (Kr)	84	83.9078	0.998902		
					- 0.0520 %	13,000
56	Barium (Ba)	138	137.920	0.999422		
					- 0.0787 %	19,675
92	Uranium (U)	235	235.0491	1.000209		
(1)	(2)	(3)	(4)	(5)	(6)	(7)

If one gram of mass were converted completely into energy, 25 million kilowatt-hours of energy would be released. Column (6) shows that there is a 0.710 % reduction of mass in converting hydrogen to helium. Hence 25 million kilowatt-hours is multiplied by 0.710 % to obtain the energy released per gram of hydrogen when hydrogen is fused to form helium. As shown in column (7), this value is 177,500 kilowatt-hours per gram.

Column (7) shows that 177,500 kilowatt hours of energy are released per gram when hydrogen is fused to form helium; 16,250 kilowatt-hours are released when helium is fused to form carbon; and 29,500 kilowatt-hours are released when carbon is fused to form iron. There are many intermediate steps involved in converting carbon into iron; this table shows the total energy released from the combined processes.

In nuclear fission, uranium-235 is split to form barium plus krypton. The writer does not know the actual isotopes of barium and krypton that are formed, and so representative isotopes are shown. There is a mass reduction of 0.0787 % in converting from uranium to barium and another 0.0520 % in converting from barium to krypton. The mass reduction in the split from uranium to barium plus krypton can be approximated by:

mass change = 0.0787 % + [84/(84 + 138)](0.0520 %) = 0.0984 %

Multiplying this by 25 million kilowatt-hours per gram shows that, in a nuclear fission bomb, 24,600 kilowatt-hours of energy are released per gram of uranium-235. In a hydrogen bomb, 175,500 kilowatt-hours are released per gram of hydrogen.

Implications of our Knowledge of the Atom

The model of the atom rests on very solid ground. Although certain details of the model are not absolutely clear, an atom definitely has a very compact nucleus, containing the protons and neutrons. The nucleus is surrounded by an electron cloud that typically has about 50,000 times the diameter of the nucleus.

A neutron star has the density of an atomic nucleus, which is about one billion tons per teaspoon. A greater density is not possible, because there is no room left in the atom for further contraction.

From the Einstein theory, modern Big Bang cosmologists have predicted singularity conditions with a density of matter that far exceeds the density of a neutron star. These singularity predictions radically violate the evidence (summarized above) that has evolved from studies of the atom. Einstein flatly rejected singularity predictions derived from his theory, because singularities contradict this physical evidence.

The 1945 quote by Einstein in Chapter 7 (p. 128) presented the case against singularities in broad and definite terms. Einstein stated that his theory would not apply under conditions of extreme density of matter. He insisted that his theory would only hold approximately in such conditions and cannot be used to predict a physical singularity.

Appendix C

Density of Matter in the Universe

C.1 Luminosity Density of the Universe

Narlikar [18] (p. 304) gives in his Eq. 9.19 the following luminosity density of the universe, derived from the *Revised Shapley-Ames Catalog*:

$$\text{Luminosity density} = 2.18 \times 10^8 \; L_{sun} h_0 / \text{Mpc}^3 \qquad \text{(C-1)}$$

where L_{sun} is the luminosity of the sun and Mpc means million parsecs, which is equal to 3.26 MLyr (3.26 million light-years). Our assumed Hubble constant H_0 is 20 km/sec per MLyr, which is 65 km/sec per Mpc. The parameter h_0 is the normalized Hubble constant, obtained by dividing H_0 by 100 km/sec per Mpc. This gives a normalized Hubble constant h_0 of 0.65. Hence Eq. C-1 can be expressed as

$$\text{Luminosity density} = 4.09 \times 10^6 \; L_{sun} / \text{MLyr}^3 \qquad \text{(C-2)}$$

C.2 Dark Matter

Measurements of the motions of galaxies and clusters of galaxies show that there must be much more dark matter (which we cannot see) than there is luminous matter (which we can see). Otherwise these rotating groups of stars would fly apart. Narlikar [18] (p. 310) gives in his Table 9.1 the data shown in the first data column of Table C-1. These values are expressed in terms of the normalized Hubble constant h_0. The last column gives the values for the ratio η that correspond to our normalized Hubble constant h_0, which is 0.65. The parameter η in Table C-1 is the ratio of total mass to luminous mass.

In Table C-1, items (1) to (4) apply to the rotational motions of galaxies. These data suggest that the total mass associated with the rotation of a single galaxy is about 8 times the luminous mass.

Items (5) to (8) involve motions of groups of galaxies. The mass ratio is much greater for a galaxy cluster than for a single galaxy. The rotation of a single galaxy involves matter in the vicinity of the galaxy, whereas the rotation of a cluster involves the total matter of the cluster.

The average distance between galaxies is about 10 million light-years (10 MLyr), and so we can allocate to each galaxy a volume of (10 MLyr)^3, which is 1000 MLyr^3. This volume is about 5 million times greater than the volume of the galaxy itself. Even though the density of dark matter is much smaller in the intergalactic space between galaxies, than in the vicinity of a galaxy, the total dark matter in intergalactic space is much greater than the dark matter close to the galaxy.

Table C-1: Average ratio (η) of total mass per luminous mass

Object	η/h_0	η
(1) Our Galaxy (inner part)	6 ± 2	3.9 ± 1.3
(2) Our Galaxy (outer part)	40 ± 30	26 ± 20
(3) Spiral galaxies	9 ± 1	5.9 ± 0.7
(4) Elliptical galaxies	10 ± 2	6.5 ± 1.3
(5) Galaxy pairs	80 ± 20	52 ± 13
(6) Local Group	160 ± 80	104 ± 52
(7) Statistics of clustering	500 ± 200	325 ± 130
(8) Abell clusters	500 ± 200	325 ± 130

Item (5) involves a pair of galaxies, and item (6) for our local group involves a small galaxy cluster. In contrast, items (7), (8) involve large clusters of galaxies, and so should give a better indication of the effects of intergalactic mass. Items (7), (8) show that the mass of intergalactic dark matter is 325 times the luminous mass of the galaxy itself. ***Therefore intergalactic dark matter (which we cannot see) is about 325 times greater than the luminous matter (which we can see).***

C.3 The Source of Dark Matter

What is the source of dark matter? Big Bang theorists are searching hard for dark matter, because their theories have difficulty explaining the early evolution of the universe unless the density of matter is close to the *critical mass density*. They have not found sufficient dark matter in the universe to achieve critical mass density.

A fundamental mistake has been made in the Big Bang search for dark matter. Silk [21] (p. 163) shows that quasar radiation has been used to measure the density of intergalactic hydrogen. Since quasars are

C. Density of Matter in the Universe   185

assumed to be at enormous distances, the density of intergalactic hydrogen is believed to be extremely small. However, Arp showed that quasars are relatively close, and so the estimates of intergalactic hydrogen derived from quasar radiation are not meaningful. Besides, most of the hydrogen in space is probably molecular hydrogen (H_2), rather than atomic hydrogen (H), and molecular hydrogen is very difficult to detect.

Dark matter probably consists primarily of hydrogen atoms in the enormous spaces between galaxies. A frantic search for exotic missing dark matter is being performed by some astronomers. This is a consequence of the lack of open debate in astronomy today. If astronomers would listen to Halton Arp, they would realize that quasars are relatively close, and so they would not use quasar radiation to measure the density of intergalactic hydrogen.

C.4 Total Mass in the Universe

For our assumed Hubble constant of 20 km/sec per MLyr, the radius r_0 of the observable universe is 15,000 MLyr (15 billion light-years). The volume of the observable Big Bang universe is

$$\text{Universe volume} = (4/3)\pi r_0^3 = 14.1 \times 10^{12} \text{ MLyr}^3 \qquad \text{(C-3)}$$

Multiplying this by the luminous density of the universe in Eq. C-2 gives the total luminosity of the universe in equivalent suns, which is

$$\text{Luminosity of universe} = 5.77 \times 10^{19} \text{ L}_{sun} \qquad \text{(C-4)}$$

Stellar mass is approximately proportional to luminosity for a large collection of stars. Hence if the sun luminosity L_{sun} is replaced by the sun mass M_{sun}, Eq. C-4 gives the approximate mass of the luminous matter. Multiplying this by 325 to account for dark matter gives the following for the total mass of the observable universe

$$\text{Mass of observable universe} = 18.8 \times 10^{21} \text{ M}_{sun} \qquad \text{(C-5)}$$

Silk [20] (p. 396) reports that the typical distance between galaxies is 10 million light-years (10 MLyr), and so the average volume of space per galaxy is 1000 MLyr^3. Dividing the universe volume in Eq. C-3 by 1000 MLyr^3 gives the following for the number of galaxies in the universe:

186 Beliefs and Facts of Creation

$$\text{Galaxies in universe} = 14.1 \times 10^9 \qquad \text{(C-6)}$$

Multiplying the luminosity density of Eq. C-2 by the average volume per galaxy (1000 MLyr^3) gives the average luminosity per galaxy:

$$\text{Luminosity per galaxy} = 4.09 \times 10^9 \, L_{sun} \qquad \text{(C-7)}$$

C.5 Mass Density of the Universe

Since stellar mass is approximately proportional to luminosity for a large collection of stars, we can replace L_{sun} by M_{sun} in Eq. C-2 to give the following for the density of luminous matter

$$\text{Density of Luminosity matter} = 4.09 \times 10^6 \, M_{sun}/\text{MLyr}^3 \qquad \text{(C-8)}$$

Multiplying this by the dark matter ratio 325 gives the total density of matter:

$$\text{Total density of matter} = 1.33 \times 10^9 \, M_{sun}/\text{MLyr}^3 \qquad \text{(C-9)}$$

Replace M_{sun} by the sun mass, 1.99×10^{33} gram. Since one light-year (Lyr) is 9.46×10^{12} km, MLyr is 9.46×10^{21} meters. This gives the following for the total mass density of the universe:

$$\text{Mass density} = 3.125 \times 10^{-24} \, \text{grams/meter}^3 \qquad \text{(C-10)}$$

The mass of a hydrogen atom is 1.67×10^{-24} gram, and so this mass density is equivalent to 1.87 hydrogen atoms per cubic meter. *This shows that our best estimate of the average mass density of the universe is 1.9 hydrogen atoms per cubic meter.*

C.6 Predicted Density of Matter

The Yilmaz cosmology model predicts an average density of matter in the universe of $(3/8\pi G T_0^2)$, where T_0 is the apparent universe age and G is Newton's gravitational constant (6.674×10^{-8} cm^3/gm-sec^2). There are 31.558 million seconds per year, and so the apparent universe age T_0 (15 billion years) is 4.734×10^{17} seconds. The above formula gives an average density of matter of 7.98×10^{-30} gm/cm^3 or 7.98×10^{-24} grams per cubic meter. A hydrogen atom has a mass of 1.67×10^{-24} gm, and so this density is equivalent to 4.78 hydrogen atoms per cubic meter.

Thus the Yilmaz theory predicts an average density of matter in the universe equivalent to 4.8 hydrogen atoms per cubic meter. When one considers the great errors involved in these calculations, this predicted density is remarkably close to the 1.9 hydrogen atoms per cubic meter density that is estimated from astronomical measurements.

Big Bang theories define a *critical mass density* for the universe. If the density of matter is less than critical, the universe should expand forever; and if the density is greater than critical the universe should eventually collapse. The critical mass density for the Big Bang theory has the same value $(3/8\pi G T_0^2)$ that is required by the Yilmaz theory, and so is also equivalent to 4.8 hydrogen atoms per cubic meter.

C.7 Theoretical Mass of the Universe

Thus, the Yilmaz cosmology model predicts a mass density of the universe of $(3/8\pi G T_0^2)$, which is also the critical mass density for the Big Bang universe. Multiply this mass density by the volume of the observable universe, $(4/3)\pi r_0^3$, where r_0 is the radius of the observable universe. This gives a total mass of $(r_0^3/2GT_0^2)$. The ratio (r_0/T_0) is equal to the speed of light c, and so the total universe mass becomes $(c^2 r_0/2G)$. The normalized relativistic mass of the sun, denoted (m_{sun}), is defined as $(M_{sun}G/c^2)$. (See, *Story* [2], p. 152.) Hence the total mass of the observable Big Bang universe for critical mass density is equal to

$$\text{Mass of observable universe} = (c^2 r_0/2G) = M_{sun}(r_0/2m_{sun}) \quad \text{(C-11)}$$

The normalized mass of the sun m_{sun} is 1.475 km. Since one light-year Lyr is 9.46×10^{12} km, the radius of the observable universe r_0, which is 15 billion light-years, is 142×10^{21} km. Substituting these values into the above equation gives 48×10^{21} M_{sun} for the universe mass.

For critical mass density, the observable Big Bang universe has 48×10^{21} times the sun mass. This is also the theoretical mass of the Yilmaz cosmology model for a sphere with a radius of 15 billion light-years. Equation C-5 gives a measured mass for the observable universe of 18.8×10^{21} times the sun mass, which is about 40 percent of the theoretical mass. When we consider the large errors involved in these calculations, this agreement between theoretical and measured values is very good.

C.8 Rate of Creation of Matter

Narlikar [18] (p. 240, Eq. 8-4) shows that the rate of mass creation to compensate for the Hubble expansion is $(3\rho/T_0)$. Setting mass density ρ equal to 4.8 hydrogen-atom/m^3 shows that 0.96 (about one) hydrogen atom is created per cubic meter every billion years, or *one hydrogen atom is created every year within a cubic kilometer.*

Our earth, with a radius of 6378 km, has a volume of 1.09×10^{12} cubic kilometers, and so 1.09×10^{12} hydrogen atoms would be created per year within a volume the size of the earth, which comes to 34,500 hydrogen atoms per second. The equivalent energy (Mc^2) of the hydrogen atom (1.67×10^{-27} kilogram) is 1.50×10^{-10} watt-second. Hence 34,500 hydrogen atoms per second is equivalent to 5.18 microwatt. *This shows that the predicted rate of creation of matter is equivalent to the continual conversion of 5 microwatts of power into matter within a volume the size of the earth.*

C.9 Size of the Initial Big Bang Universe

A star with our sun's mass but the density of water would have a radius of 780,000 km, obtained by multiplying 700,000 km (sun radius) by the cube root of 1.4 (sun density). Multiply 780,000 km by the cube root of 18.8×10^{21} (from Eq. C-5) to obtain the radius for the observable universe if it were squeezed into a single body with the density of water:

Universe radius = 20.7×10^{12} km = 2.19 Lyr   (water density)   (C-12)

One light-year (Lyr) is 9.46×10^{12} km. The density of a neutron star is 2.0×10^{14} times the density of water. The cube root of this ratio is 58,480. Hence the radius of Eq. C-12 is divided by 58,480 to obtain the following radius of the observable universe if it were squeezed to the density of a neutron-star:

Universe radius = 354 million km   (neutron-star density)   (C-13)

This is 1.55 times the radius (228 km) of the orbit of Mars around the sun. *Therefore astronomical data shows that the observable Big Bang universe would have had an initial radius 1.55 times the radius of the Mars orbit if it began with the density of a neutron star, the maximum possible density of matter.*

Appendix D

Cosmic Microwave Background Radiation

Big Bang proponents claim that cosmic microwave background radiation validates their theory. However, the Yilmaz cosmology theory also predicts this radiation. In Chapter 10, Fig. 10-1 shows that at very large distances the apparent galaxy velocity is very close to the apparent speed of light, and so the light radiated from distant galaxies should be Doppler shifted to very low frequencies. *Believe* [1] presents an analysis of this radiation in Appendix D, which is summarized in this appendix. In this appendix the plots have been modified to correspond to a Hubble constant of 20 km/sec per million light-years (rather than 25).

This cosmic microwave radiation predicted by the Yilmaz cosmology theory is equivalent to the emission from an ideal blackbody at a temperature between 2.1 and 3.4 °K. This agrees closely with the 2.73 °K blackbody temperature measured by the COBE satellite.

The general spectrum of blackbody radiation was shown in Fig. 4-2 of Chapter 4. This spectrum is expressed in terms of the half-power frequency f_h of the spectrum. The equivalent half-power wavelength L_h, is equal to c/f_h. This wavelength is related as follows to the blackbody temperature T expressed in degrees Kelvin:

$$L_h = 4.107/T \text{ millimeter (mm)}$$

The temperature of a blackbody determines the actual frequencies that it radiates, but the shape of the spectrum is the same for all temperatures.

The light radiated from our sun has a spectrum approximating that of an ideal blackbody at a temperature T of 5770 °K. The corresponding value for the wavelength L_h is 0.712 micrometers (millionths of a meter). Our analysis approximates the spectra of all stars in our universe with that of our sun.

Because the light from a galaxy is Doppler shifted toward lower frequency, the equivalent blackbody temperature of the spectrum that is received from a galaxy decreases with distance to the galaxy. Figure D-1 shows the equivalent blackbody temperature of the Doppler-shifted spectrum that is received from a galaxy at a particular true distance. This was calculated by applying the Einstein formula for Doppler frequency shift (given in *Story* [2], Appendix E) to the data in Fig. 10-1 of Chapter 10 (p. 149), which shows the apparent galaxy velocity divided by the apparent speed of light for the Yilmaz cosmology theory. It is the apparent velocity ratio that determines the Doppler frequency shift.

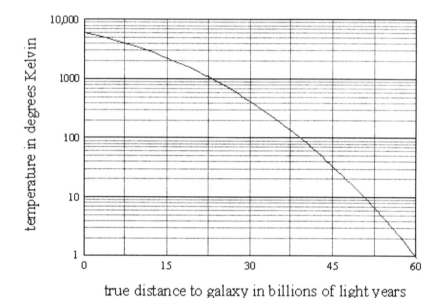

Figure D-1: Blackbody temperature of radiation received from a distant galaxy versus the true distance r to the galaxy

In Fig. D-1 the equivalent blackbody temperature decreases from 5770 °K (the blackbody temperature of the sun) for a close galaxy down to 1 °K for a galaxy at a true distance of 60 billion light-years.

Figure 10-4 on page 153 showed that the apparent density of matter should be very high as the apparent distance to a galaxy approaches the limit of 18.8 billion light-years. Because of this very high apparent density, the intensity of radiation received from a galaxy located at a large true distance should also be very large. Figure D-2 shows the photon rate that should be received per unit area of receiver surface, from galaxies at different values of true distance. The plot shows that the photon rate becomes extremely large at large true distances.

D. Cosmic Microwave Background Radiation

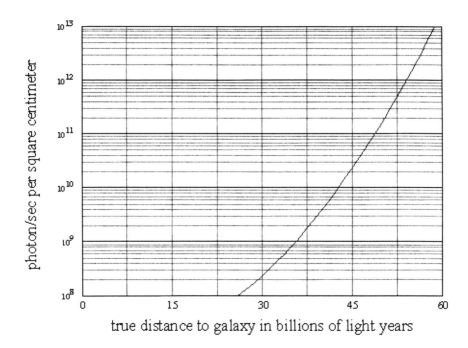

Figure D-2: Photon rate per unit area for the cosmic radiation received from galaxies at a true distance r.

By combining the data in Figs. D-1 and D-2, we can plot the intensity of the received radiation versus the equivalent blackbody temperature of the spectrum. This gives the solid curve in Fig. D-3. This shows the photon rate that would fall onto a square centimeter of receiver surface, expressed in terms of the equivalent blackbody temperature T of the received Doppler-shifted spectrum.

An ideal blackbody emits a photon rate that is proportional to the cube of the blackbody temperature. The dashed plot in Fig C-3 shows the photon rate per unit area that is emitted from an ideal blackbody.

For an ideal blackbody, the radiation is in thermal equilibrium with the molecules at the surface. We assume that this radiation level cannot be exceeded by cosmic radiation in space. If it were, the diffuse material in space should rapidly absorb the cosmic radiation. Therefore we conclude that the Doppler-shifted galaxy radiation indicated by the solid plot in Fig C-3 cannot exceed the dashed plot. This indicates that the intersection point of the two plots should give the effective blackbody temperature of the received blackbody radiation. The figure shows that this calculated temperature is 2.7 °K. Because of approximations in the analysis, there can be an error in this result. We conservatively estimate that an exact computed temperature should fall within the range from 2.1 °K to 3.4 °K.

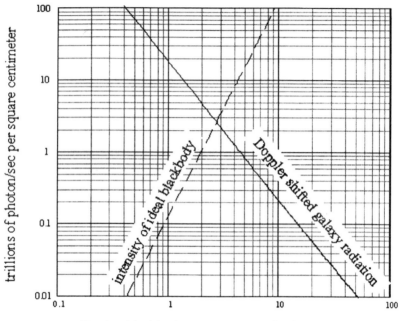

Figure D-3: Photon rate intensity of cosmic radiation received from galaxies, versus effective blackbody temperature of radiation; received radiation (solid); ideal blackbody (dashed)

Figure D-1 shows that galaxies producing blackbody radiation equivalent to a 2.7 °K temperature are at a distance of 56 billion light-years. This indicates that we should receive from galaxies at a true distance of about 56 billion light-years cosmic microwave radiation that corresponds to a blackbody at a temperature from 2.1 °K to 3.4 °K.

For comparison, the COBE satellite found that the received cosmic radiation is equivalent to the radiation from an ideal blackbody at a temperature of 2.73 °K. The COBE radiation is received with very high uniformity from all directions, which agrees with the Yilmaz cosmology theory.

Appendix E

Theoretical Discussion of Einstein and Yilmaz Theories

E.1 The Basis for the Yilmaz Theory

E.1.1 Einstein Formula for Gravitational Redshift

Chapter 6 (p. 117) showed the approximate calculation made by Einstein to determine the redshift produced by a gravitational field. He showed that when light rises through a height (h) in a gravitational field having an acceleration of gravity (g), the light experiences a redshift approximately equal to

$$\Delta L/L = gh/c^2 \tag{E-1}$$

Parameter L is the wavelength of the light, and ΔL is the wavelength increase as the light rises through the height h. The ratio $\Delta L/L$ is called the "redshift", which is the relative increase of wavelength. The redshift is the wavelength increase ΔL divided by the wavelength L.

For a spherically symmetric body of mass M and radius r, the acceleration of gravity g is equal to

$$g = GM/r^2 \tag{E-2}$$

Parameter G is the gravitational constant, which is the constant of proportionality in Newton's law of gravitational attraction. As was shown in Chapter 3, this law is expressed as

$$\textit{Gravitational force} \ = \ GM_1M_2/(d_{12})^2 \tag{E-3}$$

This is the gravitational force between two bodies of masses M_1 and M_2, separated by the distance d_{12}. Substituting Eq. E-2 into Eq. E-1 gives

$$\Delta L/L = (GM/c^2)(h/r^2) = (GM/c^2)(\Delta r/r^2) \qquad (E-4)$$

The height h has been replaced by Δr, because the height h represents a change (Δ) in the radial distance r measured from the center of the gravitational body (which could be the earth).

Calculus shows that for small changes, $(\Delta r/r^2)$ is approximately equal to $-\Delta(1/r)$, which represents the negative of the change in $1/r$. Hence the expression $-(\Delta r/r^2)$ is approximately equal to

$$-(\Delta r/r^2) = \Delta(1/r) = (1/r_2) - (1/r_1) \qquad (E-5)$$

In the elevator of Fig. 6-1 in Chapter 6, point (1) is at the floor of the elevator, point (2) is at the ceiling, r_1 is the distance of the floor from the center of the earth, and r_2 is the distance of the ceiling from the center of the earth. Substituting Eq. E-5 into Eq. E-4 gives

$$\Delta L/L = (GM/c^2)(\Delta r/r^2) = (GM/c^2)[(1/r_1) - (1/r_2)] \qquad (E-4)$$

The wavelength difference ΔL is equal to $(L_2 - L_1)$, where L_2 is the wavelength at the ceiling (point 2) and L_1 is the wavelength at the floor (point 1). The wavelength ratio $\Delta L/L$ can be approximated as $\Delta L/L_1$, which is equal to

$$\Delta L/L_1 = (L_2 - L_1)/L_1 = L_2/L_1 - 1 \qquad (E-5)$$

Combining Eqs. E-4 and E-5 gives

$$L_2/L_1 = 1 + (GM/c^2)[(1/r_1) - (1/r_2)] \qquad (E-6)$$

This gives the ratio L_2/L_1 of the wavelength L_2 of the light at the ceiling of the elevator, divided by the wavelength L_1 at the floor of the elevator. This is based on the calculation derived by Einstein showing the approximate effect of gravity on the wavelength of light.

E.1.2 Yilmaz Formula for Gravitational Redshift

Yilmaz discovered that he could solve this problem exactly. *Story* [2] (Appendix E) gives a derivation of the following exact formula that was calculated by Yilmaz:

$$L_2/L_1 = \exp\{(GM/c^2)[(1/r_1) - (1/r_2)]\} \tag{E-7}$$

The expression "exp" represents the exponential function. This is a well-known mathematical function, which can be read from a scientific pocket calculator. The function $\exp\{x\}$ is approximately equal to $(1 + x)$ if x is much less than unity. Consequently, the Yilmaz formula of Eq. E-7 closely approximates the Einstein formula of Eq. E-6 for experiments performed within our solar system, where the relativistic effects due to gravity are very small.

The frequency of an electromagnetic wave can be used as the timing reference for a clock. Consequently, this exact Yilmaz formula for wavelength given in Eq. E-7 also specifies the relativistic effect on a time measurement that is produced by gravity.

To generalize his result, Yilmaz postulated that the speed of light, measured locally in a gravitational field, is independent of the direction of the light. By combining this postulate with the exact formula of Eq. E-7, Yilmaz derived the relativistic effect produced by gravity on measurements of distance. In this manner, Yilmaz obtained rigorous formulas for the relativistic effects produced by gravity on both time and distance measurements, and thereby obtained a complete specification for the Yilmaz relativistic theory of gravity.

A detailed derivation of the Yilmaz theory is presented in *Story* [2], Appendix E.

E.2 The Einstein Gravitational Field Equation

E.2.1 Tensors of the Einstein Theory

The Einstein general theory of relativity is specified by the **Einstein gravitational field equation**, which is a tensor formula. To understand this formula, let us briefly examine the characteristics of a tensor.

Tensor theory is based on the mathematics of curved space developed by Riemann and Ricci. The most fundamental tensor of this mathematical theory is the **metric tensor**, which specifies the shortest distance between two points in curved space. The path of this shortest distance is called a **geodesic**. A geodesic in curved space is equivalent to a straight line in flat Euclidean space. A straight line is the shortest distance between two points in Euclidean space.

To illustrate the meaning of a geodesic, assume that one is flying by airplane from Boston, Massachusetts to Rome, Italy. Both cities are at approximately the same latitude, and so one might think that the airplane would fly directly east from Boston to reach Rome. However, that route is the long way. The shortest path from Boston to Rome is the **great circle route**, which can be found by passing a plane through the center of the earth, Boston, and Rome. The path where this plane intersects the curved surface of the earth is the great circle route between Boston and Rome. To follow this route, the airplane starts by flying nearly northeast, and ends the trip by flying nearly south-east.

The curved surface of the earth is 2-dimensional, and is specified by the two dimensions, latitude and longitude. A great-circle route is a geodesic expressed in 2 dimensions. The metric tensor used in Einstein's general theory of relativity is specified in 4 dimensions (3 spatial dimensions plus time). This metric tensor specifies the shortest distance between two points in 4-dimensional curved space.

The metric tensor is denoted by the form (g_{ab}), where each subscript a or b is called an "index". Each of these two indices can have one of the four values, 0, 1, 2, or 3. Index 0 specifies the time dimension, and indices 1, 2, 3 specify the 3 spatial dimensions.

The metric tensor (g_{ab}) has 4x4 or 16 separate elements. These 16 elements are arranged as follows in an array called a "matrix":

$$\begin{vmatrix} g_{00} & g_{01} & g_{02} & g_{03} \\ g_{10} & g_{11} & g_{12} & g_{13} \\ g_{20} & g_{21} & g_{22} & g_{23} \\ g_{30} & g_{31} & g_{32} & g_{33} \end{vmatrix} \qquad \text{(E-8)}$$

The tensors in Einstein's theory are symmetric, which means that the value is the same if the two indices are interchanged. For example, g_{12} is equal to g_{21}. Because of this symmetry, 6 of the 16 elements are redundant. Therefore, in the general case, each tensor of the Einstein theory has 10 independent elements.

The Einstein general theory of relativity is specified by a tensor formula called the **Einstein gravitational field equation**. Since the tensors of this formula have 10 independent elements, the Einstein gravitational field equation represents 10 independent equations.

Another tensor used in relativity theory is the **Einstein tensor**, which specifies the curvature of space, and is denoted (G_{ab}). The Riemann-Ricci mathematical theory has a complicated formula that allows one to compute the Einstein tensor (G_{ab}) from the metric tensor (g_{ab}). This is a purely mathematical theory that relates the shortest distance between two points (a geodesic) to the curvature of space. In a flat (Euclidean)

E. Discussion of Einstein and Yilmaz Theories 197

space with no curvature, all elements of the Einstein curvature tensor (G_{ab}) are zero.

As applied to the Einstein theory, 4-dimensionsl space has no curvature if there is no gravitational field nor acceleration. Within our solar system, the gravitational field is very weak, and so all elements of the Einstein curvature tensor are very small.

The final major tensor used in the Einstein theory is the energy-momentum tensor (T_{ab}). The energy-momentum tensor specifies the characteristics of energy and matter. The Einstein gravitational field equation is a tensor formula that relates the Einstein tensor (which specifies the curvature of space) to the energy-momentum tensor (which specifies the characteristics of energy and matter).

Tensors have three different forms, which are specified by the arrangement of the indices. These three forms are illustrated as follows for the Einstein tensor:

covariant tensor	G_{ab}	subscripted indices
contravariant tensor	G^{ab}	superscripted indices
mixed tensor	$G_a{}^b$	subscripted and superscripted indices

If the elements of the metric tensor are known, any other tensor can be converted from one of these forms to another by using relatively simple formulas.

The **Einstein gravitational field equation** is usually specified as follows in terms of the mixed form of the tensors:

$$G_a{}^b = -8\pi T_a{}^b \qquad (E-9)$$

Tensor $G_a{}^b$ is the mixed form of the Einstein tensor, and $T_a{}^b$ is the mixed form of the energy-momentum tensor. Since these tensors have 10 independent elements, the Einstein gravitational field equation represents 10 independent equations.

E.2.2 Solving the Einstein Equations

To apply the Einstein theory, the energy-momentum tensor is calculated for the particular physical model that is being studied. The Einstein equations are solved to determine the corresponding metric tensor. When the metric tensor is known, one can directly compute the relativistic characteristics of the physical model.

There are precise equations for calculating from a physical model the contravariant form of the energy-momentum tensor, which is denoted T^{ab}. Finding the energy-momentum tensor T^{ab} of a physical model is a straightforward calculation.

If the metric tensor is known, one can easily calculate the mixed form of the energy-momentum tensor $(T_a{}^b)$ from the contravariant form (T^{ab}). By applying the Einstein gravitational field equation of Eq. E-9, one can calculate from $T_a{}^b$ the Einstein tensor $G_a{}^b$. The Riemann-Ricci mathematical theory of curved space provides a formula for calculating the Einstein tensor from the metric tensor. This formula must be solved **backward** to determine the metric tensor (g_{ab}) that corresponds to the computed Einstein tensor $(G_a{}^b)$.

But how does one calculate $(T_a{}^b)$ from (T^{ab}) before one has calculated the metric tensor? How does one solve the very complicated Riemann-Ricci formula backward? For a general case, the Riemann-Ricci formula can yield millions of terms. These issues show that the application of the Einstein theory to a physical problem can be a daunting mathematical calculation.

This calculation is greatly simplified if a simple physical model is selected that yields a metric tensor with the following form, which is called a *diagonal metric tensor*:

$$\begin{vmatrix} g_{00} & 0 & 0 & 0 \\ 0 & g_{11} & 0 & 0 \\ 0 & 0 & g_{22} & 0 \\ 0 & 0 & 0 & g_{33} \end{vmatrix} \qquad (E\text{-}10)$$

All elements of this tensor are zero except for the four elements along the diagonal of the matrix.

Einstein could only consider very simple applications of his theory that yielded diagonal metric tensors. If the metric tensor is not diagonal, the Riemann-Ricci formula yields millions of terms, and so cannot be solved analytically. It was not until a decade after Einstein's death, when powerful computers became widely available, that the Einstein theory could be applied to physical models with metric tensors that are not diagonal.

Einstein was unable to derive an exact solution to his theory. The first exact solution was obtained by Karl Schwartzschild, who was cooperating with Einstein. Unfortunately, Karl Schwartzschild died from disease before his famous solution was printed. Einstein published the Schwartzschild solution in 1916, the same year that he presented his general theory of relativity.

For his physical model, Schwartzschild found a means of computing $(T_a{}^b)$ from (T^{ab}) without actually knowing the metric tensor (g_{ab}). To solve the Riemann-Ricci formula backward, Schwartzschild assumed a general mathematical form for the metric tensor (g_{ab}), and was able to solve the Riemann-Ricci formula for this general form.

The Schwartzschild solution was applied to calculate the relativistic effects produced by the gravitational field of the sun. From this solution, the three basic tests of general relativity were derived. As was explained in Chapter 6 (p. 119), these three tests were used to verify the Einstein general theory of relativity.

Schwartzschild had to assume a very simple physical model of our sun to obtain a solution to the Einstein theory. Our sun has 200 times the density of water at its center, and one-thousandth of the density of water at its surface. Thus, the density of the sun varies with radius by a factor of 200,000.

Nevertheless, the physical model of the sun used by Schwartzschild assumed that the sun had a constant density of matter. This was a poor representation of our sun, but was essential in order to achieve a solution to the very complicated Einstein equations.

When powerful computers became widely available in the 1960's, the equations of the Einstein theory could be solved for complicated physical models. Computers could handle the millions of terms that are produced by the Riemann-Ricci formula when the metric tensor is not diagonal. However, even with a powerful computer, solving the Einstein theory for a complicated physical model is an extremely difficult mathematical problem. A tremendous amount of research has been expended to develop computer algorithms that can achieve these solutions.

E.2.3 Limited to a Single-Body Solution

During Einstein's lifetime, solutions of the Einstein theory were limited to a single-body solution. This solution could only consider the gravitational effect of a single body, which was usually our sun. This limitation is particularly important when one considers the application of the Einstein theory to the orbit of Mercury.

As explained in Chapter 6, the axis of the elliptical orbit of Mercury rotates 1.39 arc seconds (angular-seconds) for every orbit. Since this rotation is in the direction of the Mercury motion, the orbit is said to "advance" by 1.39 arc-seconds per orbit.

Newton's theory predicts that the gravitational effects of the other planets should produce an advance of the Mercury orbit equal to 1.29 arc seconds per orbit. The Einstein theory was able to predict the remaining 0.10 arc-second advance per orbit, which could not be explained by Newton's theory. This prediction of the Einstein theory was an important step in validating the general theory of relativity.

This prediction of the Einstein theory used the Schwartzschild analysis. This was a "single-body solution", which only considered the gravitational effects of a single body, the sun. Consequently, this relativistic calculation of the Mercury orbit could only calculate the 0.10 arc second per orbit advance produced by the sun's gravity. It could not calculate the 1.29 arc second per orbit advance produced by the gravity of the other planets.

Einstein was unable to use his theory to calculate the complete (1.39 arc second per orbit) advance of the Mercury orbit, because this would have required a non-diagonal metric tensor. With an analytical calculation, he could only achieve a *single-body solution*, which could only consider the gravitational field of a single body, the sun. This analysis could not include the gravitational effects of any of the planets. Mercury was modeled as a test mass, having no gravitational field itself.

When computers became available in the 1960's to study the Einstein equations, it was generally assumed that the Einstein theory could yield multi-body solutions. However, Section E.3 will give evidence indicating that the single-body limitation still applies.

E.3 Yilmaz Gravitational Field Equation

E.3.1 Tensor for Gravitational Energy and Force

As shown in Section E.1, Yilmaz derived an exact solution to the principles of general relativity, which yielded the elements of the metric tensor. From this tensor, Yilmaz calculated the corresponding gravitational field equation, which has the following form:

$$G_a^{\ b} = -8\pi T_a^{\ b} - 2 t_a^{\ b} \tag{E-11}$$

This is similar to the Einstein gravitational field equation in Eq. E-9, except that it has the additional term $-2t_a^{\ b}$. Yilmaz calls $t_a^{\ b}$ the *stress-energy tensor for the gravitational field*. This $t_a^{\ b}$ tensor specifies the energy and force of the gravitational field.

Einstein recognized the desirability of a stress-energy tensor to describe the gravitational field. He searched for such a tensor but could not isolate a true tensor for this purpose. Einstein found a "pseudo-tensor" to describe the energy of the gravitational field. However, Einstein could not use this in his gravitational field equation, because it was not a true tensor. The Einstein pseudo-tensor is discussed by Pauli [6] (p. 176).

With the additional stress-energy tensor for the gravitational field, the gravitational field equation for the Yilmaz theory is more complicated than that for the Einstein theory. However, the Yilmaz gravitational field equation does not have to be solved, because Yilmaz has a general solution to it. Consequently, the Yilmaz theory is very much easier to apply than the Einstein theory.

The metric tensor that Yilmaz derived is diagonal, and so it consists of only four elements (g_{00}, g_{11}, g_{22}, g_{33}). For the Einstein theory, the metric tensor is diagonal only for very simple physical models. However, for the Yilmaz theory, the metric tensor is diagonal for all applications, provided the velocities are much less than the speed of light. This condition is satisfied in essentially all practical cases, and so the basic Yilmaz theory is generally more than adequate for practical applications.

Yilmaz derived a general time-varying solution to his theory that does not have a velocity limitation. However, about the only application where this general solution is needed is for characterizing gravitational waves.

E.3.2 Need for a Gravitational-Field Stress-Energy Tensor

The fact that the Einstein gravitational field equation lacks a stress-energy tensor for characterizing the gravitational field is a serious limitation of the Einstein theory. An important consequence of this weakness is that the Einstein theory cannot achieve a multi-body solution. This finding was demonstrated by an analysis performed by Professor Carroll O. Alley.

Professor Carroll O. Alley of the University of Maryland is one of the very few experts in general relativity theory who has applied his knowledge to practical applications. Alley has supervised several experiments to test the validity of predictions derived from the Einstein theory. This has included laser measurements with retro-reflectors on the moon that have allowed distances to the moon to be measured with laser beams to an accuracy of 3 centimeters. Another set of experiments measured the relativistic time delay in an atomic clock carried in an aircraft under several flight profiles. [Y10]

Alley is also intimately involved in applying general relativity corrections to the *Geophysical Positioning System (GPS)*. The GPS is an array of satellites operated by the U. S. Air Force to provide accurate position coordinates over the world for military and civilian navigation.

Alley became impressed with the Yilmaz theory and has been cooperating with Yilmaz. Alley made an important contribution to this issue with his proof that the Einstein theory cannot achieve a multi-body solution. Let us examine this astonishing result.

The Single-Body Einstein Solution. We saw in Section E.2 that Einstein could apply his theory only to simple cases that involved a single gravitational body. If the gravitational field is produced by two or more independent bodies, the metric tensor is generally not diagonal, and so an analytical solution to the Einstein equations cannot be achieved. When computers became available in the 1960's to solve the Einstein equations, it was assumed that multi-body solutions to the Einstein equations could be achieved. However, an analysis by Alley has proven that the Einstein theory is incapable of providing an interactive multi-body solution.

Alley applied the Einstein theory to a simple model consisting of a pair of infinite slabs of matter separated by a fixed distance. By assuming that the width and length of each slab was infinite, edge effects could be ignored. This resulted in a simple theoretical model to which the Einstein theory could be applied analytically. Alley found that his analysis predicts that there is no gravitational attraction between the two slabs, a result that directly contradicts experimental evidence.

Alley and Yilmaz have concluded from this analysis that the Einstein theory cannot allow an interactive multi-body solution. The Einstein theory can model multiple bodies. However, they cannot interact gravitationally, because the Einstein gravitational field equation lacks a tensor to specify the energy and force of the gravitational field. A physical discussion of this issue is given in *Story* [2], Chapter 13 (pp. 182-184).

Many scientists who have implemented computer studies of the Einstein equations object to this conclusion. They claim that they have achieved multi-body solutions in their computer studies. However, Yilmaz and Alley point out that the complicated computer algorithms that are being used are inserting results into the solutions that are not actually coming from the Einstein equations. Despite the objections to the Yilmaz and Alley conclusions, no scientist has found any flaw in the Alley analysis.

E.3.3 Variation of Speed of Light with Direction

As was shown in Chapter 8, an experimental program was partially funded by the U. S. Government to compare the Einstein and Yilmaz theories.

For tests performed within our solar system, the Einstein and Yilmaz theories yield essentially the same results. Consequently all tests that have verified the Einstein theory are consistent with the Yilmaz theory. However, there is one very sensitive test, which has been partially implemented, that could distinguish between the Einstein and Yilmaz theories. A fundamental principle of the Yilmaz theory is that the speed of light measured locally in a gravitational field is the same in all directions. The Einstein theory predicts it is different.

As explained by Prof. Carroll O. Alley [Y10], a preliminary experiment was performed under his supervision to measure the difference with direction of the speed of light between the U.S. Naval Observatory and the NASA Goddard Optical Research Facility, which are 21.5 km apart. To measure the one-way speed of light, an atomic clock was transported back and forth between the two facilities, to synchronize clocks at those facilities. The initial data were very promising, but government funding was cancelled before measurements could be made with the required accuracy to obtain definitive results.

Why was the modest funding for this very important experiment cancelled? Our government has spent hundreds of millions of dollars on other programs to study the Einstein theory.

This experiment was discussed by Ivars Peterson [42] in a 1994 *Science News* article: "A New Gravity: Challenging Einstein's general theory of relativity".

GLOSSARY

See Reference [26] for more definitions relating to astronomy.

numbers, exponential representation
 10^5 means 1 followed by 5 zeros; 10^{-5} means $1/10^5$.
 For example, $3.1 \times 10^5 = 3.1 \times 100,000 = 310,000$.
 $3.1 \times 10^{-5} = 3.1/100,000 = 0.000,031$
number prefixes:
 nano = billionth ($1/10^9$)   mega = million (10^6)
 micro = millionth ($1/10^6$)   kilo = thousand (1000)
 milli = thousandth (1/1000)
one billion = 1000 million; one trillion = 1000 billion

acceleration: The rate-of-change of velocity with time.
acceleration of gravity g: Rate of increase of velocity when a body is allowed to fall to earth; equal to 9.8 meter/second per second at earth surface.
aether (also called ether): Postulated medium that provides an absolute reference relative to which light propagates; Einstein rejected the *aether* concept.
Alpha Centauri (A and B): Pair of stars 4.3 light years away; separated by 10 to 30 times earth-orbit radius, about size of our sun; just beyond Proxima Centauri.
age of the universe, true: Big Bang theorists now generally predict a true age of the universe of about 14 billion years; the Steady State theory predicts it is infinite.
apparent age of the universe: The time since the Big Bang if the universe has always expanded uniformly at the present Hubble constant; equal to 15 billion years for a Hubble constant of 20 km/sec per million light years.
archaea: Simple cells without nuclei, but distinct from bacteria; see pp. 13-14.
Big Bang theory: The hypothesis that our universe began as a highly dense mass that exploded with a Big Bang about 15 billion years ago.
black dwarf star: The cold, dark remains of a white dwarf star after it stops radiating energy; the ultimate state of our sun.
black hole: The Einstein theory predicts that a star with a mass-to-radius ratio exceeding 240,000 times that of our sun must collapse to form a *black-hole singularity* having zero size and infinite mass density; light cannot escape from within the *event horizon* sphere that surrounds a black hole.
blueshift: Wavelength shift to shorter wavelength, defined as wavelength decrease divided by normal wavelength; proportional to star velocity toward the earth.
brown dwarf star: A dark star that has less that 1/12 of the sun mass and so is too small to ignite nuclear fusion.
celestial sphere: Ancient astronomers believed that the stars were embedded in a gigantic sphere that rotated around the earth each day; modern astronomers use the celestial sphere as an artifice to describe the apparent motion of stars caused by the earth's rotation.

Cepheid variable star: A star varying periodically in power; having a variation period that is uniquely related to its radiated power.
cosmology: The study of the evolution and large-scale structure of our universe.
diagonal elements of a tensor: Those 4 elements for which the two indices are equal. The other 12 elements are called *nondiagonal*.
Doppler wavelength shift: Wavelength change due to velocity; wavelength change divided by normal wavelength is approximately equal to radial velocity divided by the speed of light. *Story* [2], Appendix E gives exact Einstein formula.
earth parameters: diameter, 12,756 km; mean orbit radius, 149.6 million km; mass, 5.975×10^{21} metric tons; mean orbit velocity, 29.77 km/sec
Einstein gravitational field equation: Tensor formula specifying Einstein's general theory of relativity; which represents 10 independent equations.
electromagnetic field equations: Mathematical formulas derived by Maxwell, which specify the physical laws of electricity and magnetism.
electromagnetic wave: Light wave, radio wave, etc., which consists of electric and magnetic fields oscillating at right angles to one another.
equivalent energy of matter: Energy equal to Mc^2 that would be released if a mass M were converted into energy; 25 million kilowatt-hours per gram
ether: See *aether*.
eukaryote: Cell with nucleus; all multi-celled organisms consist of eukaryote cells.
event horizon: Spherical surface surrounding a *black-hole* over which speed of light is zero; Einstein theory predicts light cannot escape through this surface.
galaxy: A group of many billons of stars; nearly all of the stars of the universe are parts of galaxies; but some are parts of smaller groups called *stellar clusters*.
galaxy, types of: Milky Way galaxy, with the shape of a disk, is a *spiral galaxy*. The other major type is the *elliptical galaxy*, having only an elliptical nucleus.
gravitational constant G: The constant of proportionality in Newton's law of gravitational attraction; defined on p. 53; (6.670×10^{-8} cm^3/gm-sec^2).
Hubble constant: The ratio of receding galaxy velocity to galaxy distance for distant galaxies; the latest value is about 20 km/sec per million light years.
Hubble expansion: The 1929 finding by Hubble that galaxies are receding at velocities approximately proportional to distance.
Hubble law: The postulate that our whole universe expands at a constant rate.
Light year: Distance light travels in one year; 9.46×10^{12} kilometers.
Marmet redshift: The prediction by Paul Marmet that a hydrogen cloud produces a redshift proportional to gas density and cloud thickness.
mass, rest: Mass of object at zero velocity; see p. 114.
metric ton: 1000 kilograms, one million grams, 2205 pounds, 1.10 English tons
momentum: The product of mass times velocity.
momentum, angular: For mass elements rotating about a center of gravity, the sum (for each element) of momentum times radial distance from the center.
moon parameters: mean distance, 384,400 km, mass, 7.343×10^{19} metric ton, 1.2 % of earth mass; diameter, 27 % of earth.

North Star (Polaris): Star at end of Little Dipper handle, which is less than one degree from North Pole of Celestial Sphere; to simplify our discussion, this book treats the North Star as being exactly at the Celestial North Pole.

nebula: Originally a fixed nebulous object, including galaxies and stellar clusters; now restricted to gas clouds heated by star radiation, called *gaseous nebulae.*

neutron star: A star having the maximum mass density allowable by physical laws, consisting of tightly packed neutrons; density is 200 million tons per cubic cm.

nuclear fission: Energy released by splitting heavy atoms (uranium-235 or plutonium) used in atomic nuclear bomb and nuclear power plant.

nuclear fusion: Energy released by combining light atoms; within a star, and in a hydrogen nuclear bomb.

parallax: the relative image shift of a nearby object due to motion of the observer.

parsec: Theoretical distance of a star exhibiting an annual parallax shift of ±1 arc second because of rotation of the earth around the sun; equal to 3.26 light years.

plasma electric currents: Many hydrogen atoms in space are ionized, and the electrons produce gigantic magnetic fields that greatly affect the structure of our universe, as explained by Nobel laureate Hannes Alfven.

Proxima Centauri: Nearest star, 4.2 light years away, 1/10,000 of sun's radiation; next stars are Alpha Centauri A and B, 0.1 light year further

pulsar: A star that emits radio pulses at precise intervals, typically at 10 to 1000 pulses per second; a rapidly spinning neutron star.

quasar or quasi-stellar object (QSO): A star-like object with a very large redshift.

radial velocity: Velocity of a star, toward or away from us, along a "radial" line to the star.

red giant star: State our sun will reach in 5 billion years after converting nearly all of its hydrogen to helium; sun will swell to half the distance to the earth.

receding velocity: Velocity component away from earth.

redshift: Wavelength increase of spectral lines divided by normal wavelength.

redshift, Doppler: Redshift due to velocity; approximately equal to receding velocity divided by speed of light. (Appendix E gives exact formula.)

redshift, gravitational: Redshift produced by a gravitational field.

redshift, intrinsic: Component of galaxy or quasar redshift unrelated to velocity.

Relativity, Special theory of: Einstein's basic theory (1905), which shows how time and spatial measurements are changed by the velocity of the observer.

Relativity, General theory of: Generalization (1916) of Einstein's Relativity theory to include effects of acceleration and gravity.

Schwartzschild limit: The maximum mass-to-radius ratio of a star for which the Schwartzschild solution has a real answer; 240,000 times the ratio for our sun.

Schwartzschild solution: The first exact analytical solution (1916) derived from Einstein general relativity, specifying gravitational effects of an ideal star having constant density and no viscosity; used to verify Einstein theory.

singularity: Big Bang prediction where size shrinks nearly to zero without change of mass, so that density of matter is nearly infinite.

solar system: the planets, comets, and other bodies orbiting a star.

spectral lines: Most of the light from a star has a continuous *spectrum* like the blackbody in Fig. 4-2. A portion consists of discrete *spectral lines* produced by elements. Some are bright emission lines, and some are dark absorption lines.
spectrum: Pattern formed by passing light through a prism to separate wavelengths.
speed of ligh, c: 300,000 kilometers per second (exactly, 299,793 km/sec)..
stars, types of (See individual types)
   black dwarf, brown dwarf, neutron star, pulsar, red giant, white dwarf.
Steady State theory: In 1948 Fred Hoyle postulated that universe age is infinite, and diffuse matter is created to compensate for Hubble expansion.
stellar: pertaining to a star.
supernova, An enormous stellar explosion with temporary brightness of billions of suns; the end of life for a star with more than 8 times the sun's mass.
supernova, type 1a: A supernova that radiates a fixed peak power of about 3 billion suns, apparently caused by a white dwarf collecting matter from another star.
sun parameters: radius = 696,000 km; mass (M) = 1.989×10^{30} kg; density = 1.4; normalized mass m = 1.475 km = MG/c^2.
teaspoon, equal to 4.929 cubic centimeters.
temperature, scales of: Celsius (or Centigrade) scale is zero °C at freezing point of water and 100 °C at boiling point; Kelvin scale has Celsius intervals but is zero at absolute zero (-273.15 °C) , where random molecular motion is zero.
tensor definition: A generalized variable, usually with 16 elements in Relativity theory.
tensor, types of :
   Einstein tensor, Denoted $G_a^{\ b}$ in usual form, describes the curvature of space.
   energy-momentum tensor: Denoted $T_a^{\ b}$ or T^{ab} in usual forms, describes the properties of matter and energy.
   metric tensor: Denoted g_{ab} or g^{ab} in usual forms, describes the shortest distance between two points in curved space.
   stress-energy tensor for gravitational field: The Yilmaz tensor $t_a^{\ b}$ specifies energy and stress of gravitational field (not used in Einstein theory).
tensor, forms of: A tensor can have three separate forms; the *covariant* form has subscript indices; the *contravariant* form has superscript indices, and the *mixed* form has one subscript and one superscript index.
tensor, diagonal: This class of tensor has only 4 nonzero elements, which are on the diagonal of the tensor matrix (both indices are equal).
vector: Variable having amplitude and direction, represented by an arrow.
velocity, radial or tangential: Radial velocity is velocity in the radial direction, toward or away from earth; *tangential velocity* is perpendicular to the radius.
white dwarf: When nuclear fuel is depleted, our sun will shrink to a *white dwarf,* glowing white hot from energy released by gravity; after reaching the size of the earth, it will cool to become a *black dwarf.*
Yilmaz theory of gravity: Refinement of the Einstein general theory of relativity, which has achieved an exact solution to the principles of the Einstein theory.

BIBLIOGRAPHY

There are two sets of references. The preface Y indicates references on the Yilmaz theory.

Yilmaz Theory Bibliography

[Y1] Huseyin Yilmaz, "New Approach to General Relativity", *Physical Review*, vol. 111, No. 5, Sept. 1, 1958, pp 1417-1426,
[Y2] Huseyin Yilmaz, "New Theory of Gravitation", *Physical Review Letters*, vol. 27, No. 20, 15 Nov. 1971, pp. 1399-1402.
[Y3] Huseyin Yilmaz, "New Approach to Relativity and Gravitation", *Annals of Physics*, Academic Press, NY, 1973, pp. 179-200.
[Y4] Huseyin Yilmaz, "New Theory of Gravitation", *Proc. 4th Marcel Grossman Meeting Gen. Relativity*, Remo Ruffini, ed, Rome Univ, Italy, June 1985.
[Y5] Huseyin Yilmaz, "New Direction in Gravity Theory", *Hadronic Journal*, 1986, vol. 9 No 6, pp 281-291,.
[Y6] Huseyin Yilmaz, "Present Status of Gravity Theories", *Hadronic Journal*, 1986, vol. 9 No 6, pp 233-238.
[Y7] Huseyin Yilmaz, "Dynamics of Curved Space", *Hadronic Journal*, 1986, vol. 9 No 2, pp 55-60.
[Y8] H. Yilmaz, "Toward a Field Theory of Gravitation", *Nuovo Cimento, B Gen. Physics*, 1992, vol. 107, Iss. 8, pp 941-960.
[Y9] Yilmaz, Huseyin, "Did the Apple Fall?", in *Frontiers of Fundamental Physics*, 1994, M. Barone and F. Selleri, eds, pp. 115-124, Plenum Press, NY.
[Y10] Alley, Carroll O., "Investigation with lasers, atomic clocks [etc.] of gravitational theories of Yilmaz and Einstein", in *Frontiers of Fundamental Physics*, 1994, M. Barone and F. Selleri, eds, pp. 125-137, Plenum Press, NY.
[Y11] Huseyin Yilmaz, "Gravity and Quantum Field Theory, a Modern Synthesis", *Ann New York Acad Science*, 1995, vol 755, pp 476-499.
[Y12] Carroll O. Alley, "The Yilmaz Theory of Gravity and its Compatibility with Quantum Theory", *Ann New York Acad Science*, 1995 vol 755, pp 464-477.

General Bibliography

[1] Adrian Bjornson, *A Universe that We Can Believe*, Addison Press, Woburn, MA, 2000, ISBN 09703231-0-7.
[2] Adrian Bjornson, *The Scientific Story of Creation*, Addison Press, Woburn, MA, 2002, ISBN 09703231-2-3.
[3] Adrian Bjornson, *The Mystery of Creation*, Addison Press, Woburn, MA, 2003, ISBN 09703231-3-1.
[4] Adrian Bjornson, *Addendum to, "A Universe that We Can Believe"*, available at no cost on website www.olduniverse.com.

Books on Relativity
[5] Albert Einstein, *The Meaning of Relativity*, Princeton University Press, 5th ed., 1953, ISBN 0-691-02352-2, (1st ed. 1921), (*See* appendix for 2nd ed., 1945, p. 129).
[6] W. Pauli, *Theory of Relativity*, 1958 Pergammon Press, reprint Dover Pub, NY, ISBN 0-486-64152-X
[7] Tullio Levi-Civita, *The Absolute Differential Calculus*, 1977, Dover Pub, NY, (first Italian ed, 1923), ISBN 0-486-63401-9.

Classic Papers on Relativity
[8] Albert Einstein, "On the Electrodynamics of Moving Bodies", Annalen der Physik, 1905, vol. 17, English transl. in *The Principle of Relativity*, 1952, Dover Pub, NY, pp. 35-71.
[9] Albert Einstein, "The Foundation of the General Theory of Relativity", Annalen der Physik, vol. 49, 1916, English transl. in *The Principle of Relativity*, 1952, Dover Pub, NY, pp. 109-164.
[10] Albert Einstein, "On a stationary system with spherical symmetry consisting of many gravitating masses", *Annals of Mathematics*, Oct. 1939, vol 40, No 4, pp 922-936 (see p. 936).
[11] G. Ricci and T. Levi-Civita, "Methods de calcul differential absolu et leurs applications", *Math. Ann.*, 1901, vol. 54, pp. 125-201.
[12] J. R. Oppenheimer and H. Snyder, "On Continued Gravitational Contraction", *Physical Review*, Sept. 1939, vol 56, pp 455-459.

Books on Cosmology
[13] George Gamow, *One, Two, Three . . . Infinity*, Bantam Books, 1967, original Viking edition, 1947.
[14] Halton C. Arp, *Quasars, Redshifts, and Controversies*, 1987, Interstellar Media, Berkeley, Calif, ISBN 0-941325-00-8.
[15] Halton C. Arp, *Seeing Red*, 11998, Aperion, Montreal, Quebec, ISBN 0-9683689-0-5. (available at Internet website *www.Amazon.com*)
[16] Eric Lerner, *The Big Bang Never Happened*, Times Books div Random House, NY, 1991, ISBN 0-8129-1853-3.
[17] David Filkin, *Stephen Hawking's Universe, the Cosmos Explained*, 1997, Basic Books div Harper Collins, NY, ISBN 0-465-08199-1.
[18] J.V. Narlikar, *Introduction to Cosmology*, 1993, 2nd Ed., Cambridge U. Press, Cambridge, England, ISBN 0-521-42352-X.

[19] Fred Hoyle, Geoffrey Burbidge, and Jayant Narlikar, *A Different Approach to Cosmology,* 2000, Cambridge U. Press, England, ISBN 0-521-*66223-0*.
[20] Joseph Silk, *The Big Bang,* 1989, W. H. Freeman, NY, ISBN 0-7167-1812-X.
[21] Joseph Silk, *A Short History of the Universe,* 1994, Scientific American Library, W. H. Freeman, NY, ISBN-0-7167-5048-1.
[22] John A. Peacock, Cosmological Physics, 1999, Cambridge U. Press, United Kingdom, ISBN 0-521-42270-1.

Book on Albert Einstein
[23] Albrecht Folsing, *Albert Einstein, a Biography,* 1997, (transl. from German by Ewald Osers). Penguin Books, NY, ISBN 0-14-02.3719-4.

Books on Astronomy
[24] Arthur Berry, *A Short History of Astronomy, from Earliest Times through the Nineteenth Century,* first published in 1898, reprinted in 1961 by Dover Publications, NY.
[25] Donald Goldsmith, *The Astronomers,* 1991, St. Martin Press, NY, ISBN 0-312-05380-0.
[26] Kevin Krisciunas and Bill Yenne, *The Pictorial Atlas of the Universe,* 1989, Mallard Press, ISBN 0-792-45200-3.
[27] Valerie Illingworth & John Clark, *Dictionary of Astronomy,* 4th ed, Checkmark Books, NY, ISBN 0-8160-4284-5.
[28] Terrence Dickinson, *The Universe and Beyond,* 3rd ed, 1999, Firefly Books Ltd., Ontario, ISBN 1-55209-361-1.

Books on Physics
[29] Isaac Newton, *The Principia,* English transl. by Andrew Motte, 1995, Prometheus Books, ISBN 0-87975-980-0.
[30] Isaac Newton, *Opticks,* Dover Pub, NY, 1952, Library Congress, 52-12165.
[31] Isaac Asimov, *Understanding Physics: The Electron, Proton, and Neutron,* 1966, Signet Books, New American Lib., NY, Lib. Congress 66-17227.
[32] Ronald Clark, *The Birth of the Bomb,* 1961, Horizon Press, NY, Lib. Cong. 61-15338.
[33] Dwight C. Gray, *American Institute of Physics Handbook,* 2nd Edition, McGraw-Hill, NY, 1963 (pp. 7-9 to 7-12).

Papers on Astronomy and Cosmology
[34] Geoffrey Burbidge, "Why Only One Big Bang", *Scientific American,* February, 1992, page 120.
[35] Peebles, Schramm, Turner, and Kron, "The Evolution of the Universe", *Scientific American,* Oct. 1994, pp. 53-65.
[36] Ann Finkbeiner, Astronomy: Hubble Telescope Settles Cosmic Distance Debate, or Does it?", *Science,* May 28, 1999.
[37] Paul Marmet, "A New Non-Doppler Redshift", presented in Internet website: www.newtonphysics.on.ca.
[38] L. Doyle, H. Deeg, and T. Brown, "Searching for Shadows of Other Planets", *Scientific American,* Sept. 2000, pp. 58-65.

[39] O. Baker, "Planetary potential surrounds most stars", *Science News*, Oct. 9, 1999, vol. 156, No. 15, p. 231.
[40] Jesse L. Greenstein and Maarten Schmidt, "The Quasi-Stellar Radio Sources 3C 48 and 3C 273", *Astrophysical Journal*, vol. 140, No. 1, July. 1964, pp. 1-34.
[41] "Plethora of quasars", *Science News*, Jan 23, 1999, p. 57, vol. 155, No. 4.
[42] Ivars Peterson, "A New Gravity: Challenging Einstein's general theory of relativity", *Science News*, Dec. 3, 1994, vol. 146, pp 376-378.
[43] Bruce Balick and Adam Frank, "The Extraordinary Death of Ordinary Stars", *Scientific American*, July 2004, pp 51-59.

Creation of Life on Earth
[44] James F. Keating, "When Methane Made Climate", *Scientific American*, July 2004, pp 78-85.
[45] "Earliest Evidence of Complex Life", *Science News*, Aug. 28, 1999, vol 156, no 9, p 141.
[46] J. Madeleine Nash, "When life exploded", *Time*, Dec. 4, 1995, vol 146, no 23, pp 66-74.
[47] Richard Monastersky and O. Louis Mazzatenta, "Life Grows Up", *National Geographic*, April 1998, vol 193, no 4, pp 100-115.
[48] Paul F. Hoffman and Daniel P. Schrag, "Snowball earth", *Scientific American*, Jan. 2000, vol 282, no 1, pp 68-75.
[49] R. Monastersky, "Waking Up to the Dawn of Vertebrates", *Science News*, Nov. 6, 1999, vol 156, no 19, p 292.
[50] Kerri Westenberg, "From Fins to Feet", *National Geographic*, May 1999, pp. 115-126.
[51] Richard Monastersky, "Out of the swamps", *Science News*, May 22, 1999, vol 155, no 21, pp 328-330.
[52] Michael J. Benton, *The Reign of the Reptiles*, 1990, Crescent Books, NY, ISBN 0-517-02557-4.
[53] Michael J. Benton, *The Rise of the Mammals*, 1991, Crescent Books, NY, ISBN 0-517-02561-2.
[54] William K. Hartmann and Ron Miller, *The History of Earth*, 1991, Workman Pub, NY, ISBN 0-89480-756-0, p. 176.
[55] *Atlas of Life on Earth*, Barnes and Noble Books, 2001, ISBN 0-7607-1957-8.
[56] Peter Ward, "The greenhouse extinction", *Discover*, Aug, 1, 1998, vol 19, pp 54-55; and, "Mass extinction: the big heat", *The Economist*, Aug. 28, 1999, vol 352.
[57] Christopher P. Sloan and J. Louis Mazzatenta, "Feathers for T. Rex", *National Geographic*, Nov. 1999, vol 146, no 5, pp 98-107.
[58] D. M. McLean, "A terminal Mesozoic 'greenhouse': lessons from the past", *Science*, 1978, vol 201, pp 401-406.
[59] Dewey McLean, "Dinosaur Volcano Greenhouse Extinction", May 2000, Virginia Polytechnic Institute website, www.fbox.vt.edu, search for "Dewey McLean".
[60] Richard Leakey and Roger Lewin, *Origins Reconsidered*, 1992, Doubleday, NY, ISBN 385-46792-3.
[61] Roger Lewin, *The Origin of Modern Humans*, 1993, Scientific American Library, W. H. Freeman, NY, ISSN 1040-3213 (Ch. 1).
[62] Ian Tattersall, "How We Came to Be Human", *Scientific American*, Dec. 2001, pp. 56-63; and *The Monkey in the Mirror: Essays on the Science of What Makes Us Human*, Harcourt, 2002.

[63] Roy J. Britten, "Divergence between Samples of Chimpanzee and Human DNA Sequences is 5%, Counting Intels", *Proc. National Academy of Sciences*, October 15, 2002, pp. 13633-13635, v. 99.
[64] Stephen Freeland and Laurence Hurst, "Evolution Encoded", *Scientific American*, April 2004, pp 84-91.
[65] Stanley Prusiner, "Detecting Mad Cow Disease", *Scientific American*, July 2004, pp 86-93.

References to Scientific History

[66] Will Durant, *The Reformation*, 1957, MJF Books, NY, ISBN 1-56731-017-6, pp. 855-863.
[67] Will and Ariel Durant, *The Age of Reason Begins*, 1961, MJF Books, NY, ISBN 1-56731-018-4, pp. 584-611.
[68] Will and Ariel Durant, *The Age of Louis XIV*, 1963, MJF Books, NY, ISBN 1-56731-019-2, pp. 531-547.
[69] Hal Hellman, *Great Feuds in Science*, 1998, John Wiley, NY, ISBN 0-471-16980-3, Ch. 1.
[70] I. B. Cohen, *Arch. Intern. d'Histoire des Sciences*, Oct-Dec. 1958, vol 11.
[71] David Halliday and Robert Resnick, *Physics*, 1966, John Wiley, NY, Library Congress, 66-11527 (p. 1074).

Additional Reference

[72] David Morrison, *Exploring Planetary Worlds*, Scientific American Library, Freeman, 1993, ISBN 0-7167-5043-0.

INDEX

Numbers in brackets [ ] are Bibliography references.

absolute differential calculus, 119
acceleration, 51-53,55-56,114-118,120, 131,193,197
acceleration of gravity, 51,55-56,115, 193
aether, 105-107
age, 6, 11,15-18, 27, 35, 90-94, 119, 168, 186, xiv
age dilemma, 93-94, 168
age, ice, 17-19
age of earth, 6, xiv
age of mammals, 15, 17, 27, 35, xiv
age of man, 11
age of reptiles/dinosaurs, 15, 16, 27
age of sun, 6
age of stars, 93,168
age of universe, 90, 91, 93, 186
 see also, universe age
Alfven, Hannes, 66-67,138, 144-147, 162, [16]
algae, 15-16, 22-23
Alpher, Ralph, 99
Alley, Carroll O, 136, 201-203
amphibian, 7, 15-16, 20-21, 23, 35, 159, xiv
Andromeda M31 galaxy, 87
angular momentum, 64-67, 144, 161-162
apparent galaxy distance, 151, 153, 157
apparent galaxy velocity, 149, 157

apparent limit of universe, 153, 157
archaea, 7, 13-14, 18
Aristotle, 42, 45
Arp, Halton, 138-144, 185, xvii, [14, 15]
asteroid, 61
Azimov, Isaac, 174, [31]

bacteria, 4, 7, 10, 13-16, 18, 23, 179
Berry, Arthur, 44, [24]
big bang theory, 90-101, 124,129-130, 132, 134, 137-140, 144-147, 149-151, 158, 163-170, 174, 182, 184-185, 187-189, xiii, xv-xvi, xviii
birds, 2, 8, 15-17, 20, 22-27, 159
blackbody radiation, 74-76, 98-100, 158, 165, 170, 189-199
 see also, cosmic microwave radiation
black hole, 96, 124, 126-127, 129-130, 132-133, 137, 166-168. 170, xviii
 Einstein refutation of, 124-127
 event horizon, *see,* event horizon
Bondi, Hermann, 91, 92, 100
Brahe, Tycho, 47
Burbidge, Geoffrey, 100, 138-140, 142, [19, 34]

calculus, 51-53, 57, 119, 194
calculus, absolute, 119
calendar, Julian & Gregorian, 40-41
Cavendish, Henry, 53-54

celestial sphere, 36-40, 42
  coordinates, 37-38
  ancient theory of, 42
Cepheid variable stars, 77, 87, 89
clock rate, change of,
  special relativity, 109-112
  general relativity, 117, 120-123, 134
  Yilmaz cosmology theory, 150
  Yilmaz gravity theory, 134
comet belts, 64
Copernicus, 42, 46-47, 49
cosmology theories,
  see universe, models of
cosmic background explorer satellite,
  COBE, 99-100, 158, 165, 170, 189, 192
cosmic microwave radiation,
  95, 98-100, 158, 165-166, 168-170, 189-192
  see also blackbody radiation
  big bang prediction of, 98-100
  Yilmaz prediction, 100, 158, 189-192
  Bell Labs measurement, 98-99
  COBE satellite measurement,
    99-100, 158, 165, 170, 189, 192
creation of
  matter, 163-164
  earth, 5-7, 160-162
  life, 7-8, 13-35, 160
  solar system, 5, 64-67
  stars, 77-83, 161-162
  sun, 3, 77-80, 161-162
  universe, 163-170
curvature of space, 118-119

dark matter, see matter, dark
death of sun, 78-79
density of universe,
  see universe mass density
density of matter, see matter, density
Dicke, Robert, 98
Dickinson, Terrance, 77, 97, 127, 166, [28]
dinosaur, 8, 15-17, 22-27, 61, xiv

distances to stars and galaxies,
  values of, 68-70, 87-90, 149-152
  measurement of, 72-77, 87-89
  apparent, 151, 153, 157
  true, 150, 151, 155
Doppler wavelength shift, 86-87, 140-141, 157, 189-191
dwarf star,
  white, 78-79, 82-83, 89
  black, 79, 156

earth, creation, 5-7, 64-67
earth, spherical, 42-43, 45-46
ecliptic, 38-40
Einstein, Albert, 70, 92, 96, 98, 101, 107-119, 112-116, 118-137, 144, 146, 150, 152, 154, 157-158, 166-169, 182, 190, 193--203, xiii, xvi, xviii-xix, [5, 8, 9, 10]
Einstein general relativity properties
  arbitrariness of, 136-137
  clock rate, 117, 120-123
  computer studies of, 123
  curvature of space, 118-119
  singularity predictions, 124-128
  pseudo-tensor, 201
  single-body solution, 199-200, 202-203
  spatial contraction, 118, 121-123
  speed of light, 118, 121-123
  tensors, 195-199
  unified field theory, 135-136
  verification of, 120
Einstein gravitational field equation,
  see gravitational field equation
Einstein photo-electric effect, 108
Einstein rejection of singularities, 124-128
Einstein relativity theories
  Einstein general relativity, 114-123
    129-131, 136-137, 152, 199-202, xviii-xix
  Einstein special relativity, 102-114
epicycle, 43-44, 47
equinox, vernal & autumnal, 39-41

Index   215

electron, 66, 80-83, 102-104, 114, 162, 174-177, 182
electromagnetic wave, 102-105, 195
electric field, 102-104
electromagnetic field equation, 104
elements in universe, 86
energy-plus-matter conservation, 156
energy-to-matter conversion, 113
equivalence, acceleration and gravity, 115-116
ether, *see* aether
eukaryote, 14-15, 23
event horizon, 126-127

field equation, gravitational,
  see gravitational field equation
field equation, electromagnetic,
  see electromagnetic field equation
Filkin, David, 96, 167, [17]
fish, 7, 15-16, 19-21, 23, 26, 35, 160, xiv
Fischer, J. R., 93-94
Folsing, 128, [23]
fourth dimension, 112-113
Freedman, Wendy, 89
Fresnel, Augustin, 105
fission, nuclear, 3-4, 125, 179-182
fusion, nuclear, 3-4, 6, 77-79, 124, 161, 180-181

galaxies,
  M31 and M33, 87
  M51 Whirlpool, 2, 85-86
galaxy, Milky Way, 2, 5, 66-67, 70-71, 79-80, 85-86, 93, 141, 161, 168, xv
galaxy distance
  apparent, 151, 153, 157
  true, 150, 151, 155
galaxy velocity
  apparent, 149, 157
  true, 155
Galileo, 47-52, 58, 145-146
Gamow, George, 92, 94-99, 101, 127, 164-165, xv-xvi, [13]
Geller, Margaret, 94

geodesic, 119, 159, 195-196
geodesic equations, 159
geological period,
  Cambrian, 7, 15-19, 24, 35, 160
  Carboniferous, 22, 35
  Cretaceous, 22, 25-27, 35
  Ediacara, 17
  Jurassic, 25, 35
  Permian, 24-27, 35
  Triassic, 24-26, 35
Gold, Thomas, 91-92, 100
Goldsmith, 124, [25]
gravity, acceleration of,
  see, acceleration of gravity
gravity wave, *see,* wave, gravitational
gravitational constant G, 53-54, 186, 193
gravitational field equation,
  Einstein, 119, 129-130, 132-135, 146, 152, 154, 169, 195-198, 200-202, xviii, xix
  Yilmaz, 130, 132, 135, 200-201, xviii
gravitational redshift,
  see, redshift, gravitational
gravitational theories,
  Einstein, see Einstein gen. relativity
  Newton, *see,* Newton gravity theory
  Yilmaz, *see,* Yilmaz gravity theory
Greenstein, Jesse L., 140-141, [40]

Hawking, Stephen, 96, 124, 167-168
Haynes, Margaret, 94
helium, 3-5, 72, 78-79, 81, 113, 161, 175, 177, 180-182
Herman, Robert, 99
Hertz, Heinrich, 104
Hipparchus, 43
Hoyle, Fred, 91-92, 100, 142, 154, 158, 165-166, 169, [19]
Hubble, Edwin, 85, 87-91, 93-94, 100, 141, 143, 148-149, 152, 154, 157-158, 163-164, 169, 183, 185, 188-189

Hubble constant, 88-91, 93, 141, 148, 152, 183, 185, 189
Hubble expansion, 85, 88, 90-91, 94, 100, 143, 148, 157-158, 163, 169, 188
Hubble expansion rate, 148
Hubble law, 89, 149, 155, 157
Hubble Space Telescope, 73, 89
Huchra, John P., 94
human evolution theories
   language, 30-32
   multi-regional, 33
   replacement, 33
humans, related species,
   australopithecus, 8, 27-28
   homo erectus, 8, 28-33
   homo habilis, 8, 28-29
   homo sapiens, archaic, 30
   modern humans, 11-12, 29-35
   Neanderthal man, 11-12, 30, 32-34
hydrogen, 3-7, 13, 53, 72, 78-81, 86, 88, 91, 104, 113, 143-144, 148, 156, 161-164, 169, 175-176, 178-182, 184-188, xv

ice age, 17-19
indices in tensors, 196-197
infinite mass density, *see* singularity

Jupiter, 5, 42, 58-63

Kepler, Johannes, 47-48, 50-52, 146

language development, 12, 30-32
Leavitt, Henrietta, 77
Lerner, Eric, 66, 93-94, 99, 138, 144-147
Leibniz, G. Wilhelm, 57
Levi-Civita, Tullio, 119, [7, 11]
Lewin, Roger, 11, [60, 61]
life on earth, 13-35
life in universe, 67-72
light rays, bending, 120

light, speed of,
   measurement of, 108-109
   constancy of, 106-107
   variation with acceleration, 118
   variation with gravity, 118, 121-123, 133, 150
light, theory of,
   aether concept, 105
   corpuscular, 105
   electromagnetic, 102-105
   photon, 108
   wave, 105

Magellanic clouds, 77
magnetic field, 102-104
mammals, 3, 15-17, 20, 23-27, 35, 160, xiv
Marconi, Guglielmo, 104
Marmet, Paul, 90-91, 143, [37]
Marmet redshift, *See*, redshift, Marmet effect
Mars, 5, 42, 58-63
mass, rest, 113-114
mass, variation with velocity, 113-114
mass-energy conversion, 113
mass density, 92, 126, 184, 186-188
   see also, matter, density of
   critical mass density, 184, 187
   Yilmaz prediction, 148, 186-188
matrix, 196, 198
matter, dark, 95, 148, 183-186, xvi
matter, density of, 84, 91-92, 94, 96-97, 101, 125-128, 148, 151-152, 154, 158, 164, 166-169, 177, 182-184, 186-188, 190, 199, xv-xvi, xix
   see also, mass density
matter and energy conservation, 156
   matter-to-energy conversion, 113
Maxwell, J. Clerk, 104-105
measurement for star and galaxy
   distance, 72-77, 87-89
   velocity, 86-87
Messier, Charles, 85-87

Mercury, 5, 42, 46, 58-63, 120, 199-200
   relativistic effect, 61, 120, 199-200
meteorite, 6-8, 13-18, 24, 26-27
metric tensor, See tensor, metric
Milky Way, see galaxy, Milky Way
momentum, 137
momentum, angular, 64-67, 144, 161-162
multi-body solution
   of Einstein theory, 199-200, 202-205
myth and cosmology, 145-147

Narlikar, Jayant V., 100, 142, 183, 188, [18, 19]
nebula, 2, 78, 84-87, 91, 143
nebula, planetary, 78, 80
Neptune, 5, 42, 58-63
neutrino, 80, 174
neutron, 80-84, 174-182
neutron star, 82-84, 92, 101, 124-125, 127, 133-134, 156, 164, 167
Newton, Isaac, 51-55, 57, 105, 119-120, 159, 186, 193, 200, [29, 30]
Newton's gravity theory, 51-55, 57, 119-120, 159, 186, 193-194, 200
Newton's optics research, 51, 59, 105

observable universe, 89, 95, 97-98, 103-104, 151, 169, 185, 187-188, xv, xvi
Oppenheimer, J. Robert., 124-127, [12]

parallax, 72-73, 77
parsec, 73, 89, 103
Pauli, W., 83, 201, [6]
Pauli exclusion principle, 83
Peacock, John A., 129, [22]
Peebles, James, 97, 99, 145, 166, xvi, [35]
Penrose, Roger, 96, 124, 167-168
Penzias, Arno, 98-100, 104
photoelectric effect, 108
photosynthesis, 7, 13-16, 18, 23

photon, 108, 190-191
plants, terrestrial, 7-8, 13, 15-16, 18, 21-23. 28, 68
plasma,, 66-67, 138, 144-147, 162
Pluto, 5, 42, 58-63
proton, 80-83, 97, 101, 135, 162, 166, 170, 174-178, 181-182, xvi
Proxima Centauri, 69-70
pseudo-tensor, Einstein, 201
pterosaur, 25-27
Ptolemy's Almagest, 43-45, 47
pulsar, 84
Pythagoras, 42, 45

quantum mechanics, 83, 135-136, 176
quasar, 91, 138, 140-144, 162, 184-185, xvii

radiation, blackbody,
   see blackbody radiation
radio wave, 67, 84, 102, 104, 140
redshift, 86-87, 90-91, 93, 117, 120, 131, 140-144, 157, 163, 193-194
   velocity (Doppler), 86-87, 93, 117, 140, 162
   gravitational, 117, 120, 131, 144, 193-194
   intrinsic, 143-144
   Marmet effect, 90-91, 143-144
relativistic effects
   clock rate, 110, 117-118, 120-123, 134, 150
   spatial contraction, 110, 118, 121-123, 134, 150
   speed of light, 106-107, 118, 121-123, 133, 150
   synchronization, 111
   wavelength, 116-117, 144
relativistic effects, gravitational
   clock rate, 117-118, 121-123, 132-134, 150
   Hubble expansion, 149, 155
   spatial contraction, 118, 121-123, 132-134, 150

speed of light, 118, 121-123,
   132-134, 150
 wavelength, 116
relativistic effects, velocity
 clock rate, 110-112
 spatial contraction, 110-112
 simultaneity, 111-113
relativity theory, *see*
 Einstein relativity theory
 Yilmaz gravity theory
reptiles, 7-8, 15-16, 20-21-21, 23-27, 160
 Diapsid, 23-24, 26
 Synapsid, 23-24, 26, 35
Ricci, Gregorio, 118-119, [7, 11]
Riemann, Bernhard, 118-119
Sandage, Alan, 89
satellites of planets, 63
Saturn, 5, 42, 58-63
Schmidt, Maarten, 140-141, [40]
Schwartzschild, Karl, 119-123, 125-127, 133, 150, 168, 198-200
Schwartzschild limit, 122-126, 133, 150
Schwartzschild singularity, 125, 127
Silk, Joseph, 84, 97, 145, 184-185, [20, 21]
single body solution of Einstein theory, *see*, multi-body solution
singularity, black hole & big bang, 92, 96-98, 101, 124-130, 132-134, 166-168, 170-171, 174, 182, xvi-xix
simultaneous events, 111
Snyder, H., 125-127, [12]
solar system, *see*, creation of
solstice, summer & winter, 38-40
sound,
 speed of, 105-106
 wave concept, 102, 105
spatial contraction,
 from gravity, 118, 121-123, 134
 from velocity, 110-112
 Yilmaz cosmology model, 150-153
speed of light, *see* light, speed of

spectral/spectrum, 51, 57, 73-77, 86-87, 90, 98-99, 107, 140, 165, 176, 189-191, xvii
spectrum, blackbody, 75-76, 99
stars
 age of, 93, 167
 distance measurement, 72-77
 great distance to, 68-72
 velocity measurement, 86-87
stars, variable
 Cepheid, 77, 87-89
 RR Lyrae, 87
 quasar variation, 141
steady state theory, 90-92, 100, 134, 154, 157-158, 163-166, 169-170, xviii
stress-energy tensor, *see* tensor type
string theory, 135-136
sun
 characteristics, 59, 62
 creation of, 3-5, 64-67, 161-163
 life and death of, 77-79
supernova, 79-80, 83-84, 89, 96, 124, 142, 157, 162
synchronized, 70, 109-112

Tattersall, Ian, 12, 31-32. 34, [62]
telescope, 48-51, 58, 73, 78, 85, 87, 89, 95, 140, 143, xv, xvii
temperature, 3, 4, 6, 14, 18, 24, 25, 63, 68, 74-79, 98-100, 158, 161, 165, 170, 174, 180, 189-192
temperature, blackbody, 74-77
temperature, stellar, 74, 77
tensor, 118-119, 129, 132, 194-204
tensor analysis, 118
tensor, diagonal, 198
tensor explanation, 195-199
tensor, types of
 Einstein, 196-197
 energy-momentum, 197-198
 metric, 195-196
 pseudo-tensor, 201
 stress-energy for gravity, 200-201

tensor forms of, 197
   (contravariant, covariant, mixed)
Tulley, Brent, 93-94

universe age dilemma, 93-94, 168
universe age
   apparent, 90, 93, 186
   Big Bang, 90, 93
   ages of oldest stars, 93
   Steady-state prediction, 91, 169
universe, mass density of,
   see mass density
universe, models of,
   big bang, *see* big bang theory
   expansion is *Apparent*, 90-91, 163-164
   steady-state, *see* steady state theory
universe radius, big bang,
   (observable universe), 89-90
universe radius, Yilmaz, 150-153
universe, structure of,
   (filament and ribbon), 93-94
unified field theory, 135-136
Uranus, 5, 42, 58-63

vector, 54-55
Venus, 5, 42, 46, 58-63
vertebrate, 7, 9, 15-16, 19-21, 35, 160

wave, electromagnetic, 102-105, 195
wave, gravitational, 132, 156, 201
wave, light, 102, 104-106, 156
wave, radio, 102, 104
wave, sound, 102, 105
wave, water, 102

Wilson, Robert, 98-100, 104
Wolpoff, Milford, 33

Yilmaz, Huseyin, 92, 100, 129-137, 144, 148-152, 154, 156-158, 166, 169-170, 186-187, 189-190, 192-195, 200-203, xviii-xix
Yilmaz cosmology theory.
   clock rate, 150
   creation of matter, 154, 156, 188
   cosmic microwave radiation, 158, 189-192
   Hubble expansion, 149, 154-159
   matter in universe, 186-187
   spatial contraction, 150-153
   speed of light, 150
   uniqueness, 152
   universe density, 148
Yilmaz gravity theory,
   derivation, 131-132, 193-195
   clock rate, 134, 150
   gravitational field equation, 132
   gravity stress-energy tensor, 132, 200-203
   spatial contraction, 134, 150
   speed of light, 133, 150
   singularity, lack of, 132-134
   wavelength, 131
   quantum mechanics, 135-136
   uniqueness of, 136-137, 152
   time-varying solution, 132
Young, Thomas, 105